Strategies of U.S. Independent Oil Companies Abroad

Research for Business Decisions, No. 13

Other Titles in This Series

No. 11 Commercial Banks and the Creditworthiness of Less Developed Countries	Yoon-Dae Euh
No. 12 Growth, Profit and Long Range Planning in Banks	Hans E. Klein
No. 14 Bank Credit Cards for EFTS: A Cost-Benefit Analysis	Robert W. McLeod
No. 15 Uses of Financial Information in Bank Lending Decisions	Ray G. Stephens
No. 16 Understanding Personality and Motives of Women Managers	Harish C. Tewari
No. 17 Plant Location Decisions of Foreign Manufacturing Investors in the U.S.	Hsin-Min Tong

Strategies of U.S. Independent Oil Companies Abroad

by
Jean-Francois G. Landeau

Produced and distributed by
University Microfilms International
Ann Arbor, Michigan 48106

Library of Congress Cataloging in Publication Data

Landeau, Jean Francois, 1944-
Strategies of U. S. independent oil companies abroad.

(Research for business decisions ; no. 13)
Originally presented as the author's thesis, Harvard.
Bibliography: p.
Includes index.
1. Petroleum industry and trade—United States.
2. Investments, American. I. Title. II. Series.

HD9565.6.L35 1979 332.6'7373 79-22115
ISBN 0-8357-1052-1

To my wife, Marie-Noëlle

CONTENTS

ACKNOWLEDGMENTS xi

CHAPTERS

I. INTRODUCTION 1

I. Aims of the Study 1
II. The Sample Definition 1
III. Overview on Methodology 6
IV. Organization of the Manuscript 8

PART I THE FIRST DECISION TO INVEST ABROAD

II. THEORIES AND FACTS ON THE INDEPENDENTS' FIRST ENTRIES ABROAD AFTER WORLD WAR II 13

Introduction 13
I. The International Business Literature on Foreign Direct Investment 13
II. Literature on the Independent Oil Companies 15
III. The Conditions Surrounding the Independents' Initial Entries Abroad 15
IV. Companies' Justifications for Going Abroad 25
V. Formulation of a Hypothesis About the Independents' Initial Entries Abroad After World War II 25

III. TEST OF A HYPOTHESIS ABOUT SOME INDEPENDENTS' FIRST ENTRIES ABROAD AFTER WORLD WAR II 33

Introduction 33
I. Average Year of Entry Abroad 33
II. The Evidence of Threat by Competitors 34
III. The Influence of Size 36
IV. The Influence of the Vertical Integration's Balance 40
V. Influence of International Experience Prior to World War II 41
VI. Regression Test 41
Conclusion 45

CONTENTS

PART II THE SUBSEQUENT DECISIONS TO INVEST AND DIVEST

IV. A FRAMEWORK FOR THE ANALYSIS OF INTERNATIONAL STRATEGIES OVER TIME 51

Introduction 51
I. Business Policy Literature on Strategy 51
II. International Business Literature on International Business Strategy 52
III. Industrial Organization Literature on Strategic Groups 53
IV. The Conceptual Framework That Will Be Used 54

V. THE MODEL OF "MULTINATIONALIZATION THROUGH VERTICAL INTEGRATION" TESTED 63

Introduction 63
I. The Model Applied to the Pre-World War II Period 63
II. The Model Applied to the Post-World War II Period 67
III. Influence of Pre-World War II International Experience on Post-World War II International Strategy 71

VI. STRATEGIC GROUP I: EXPLORERS-PRODUCERS 75

Introduction 75
I. Comparative Analysis of the Strategic Group's Members 75
II. A Case Study: Amerada/Amerada-Hess 82
Conclusion 90

VII. STRATEGIC GROUP II: TOTALLY VERTICALLY INTEGRATED COMPANIES 93

Introduction 93
I. Comparative Analysis of the Totally Integrated Independents' International Strategies Between 1945-1976 93
II. A Case Study: Atlantic/Atlantic Richfield 104
Conclusion 114

CONTENTS

VIII. CONCLUSION 117

I. Summary of the Main Findings 117
II. Implications for the Oil Industry 118
III. Contribution to Existing Bodies of Knowledge and Further Lines of Research 120

APPENDICES

1. SELECTED BIBLIOGRAPHY ON INTERNATIONAL PETROLEUM 123

2. DATA ABOUT THE INDEPENDENTS' INITIAL ENTRIES ABROAD AFTER WORLD WAR II 127

3. VARIANTS TO THE BASE REGRESSION TEST 129

4. INDEPENDENTS' EXPERIENCES ABROAD BEFORE WORLD WAR II 133

5. INTERACTION BETWEEN INDEPENDENTS IN FOREIGN EXPLORATION 145

INDEX 151

ACKNOWLEDGEMENTS

I would like to thank the members of my Thesis Committee, Professors Alfred D. Chandler, Raymond Vernon and Robert B. Stobaugh as Chairman, for offering sharp comments and helpful suggestions at the critical stages of my work. Without them this thesis would not be.

I also wish to thank other faculty members of Harvard Business School, Joseph L. Bower, Mel Horwitch, Frederick T. Knickerbocker, Malcom Salter, Walter H. Vandaele, and Louis T. Wells, as well as my colleague Alfred Sant, whose comments on my research proposal helped me to sharpen it.

A special thanks goes to Professor James Baughman who, as Director of the Doctoral Program, was key in this enterprise. His advice has always been most valuable and I am glad to have followed it.

I want to acknowledge my debt to Harvard Business School for its generous support through fellowships and loans. I am grateful to the Division of Research for granting me a one-year thesis fellowship and to the Energy Project, directed by Professor Stobaugh, for providing financial support for the field research and the materialization of this thesis.

I want to thank all the managers of the oil companies I visited who gave me their time.

CHAPTER I

INTRODUCTION

I. Aims of the Study

The basic purpose of this thesis is to examine strategy over time: not just to compare several companies' strategic positions every five or ten years, but to take a dynamic look at a continuum of strategic outcomes over a long period of time. The objective is to suggest answers to the questions: What does strategy look like over a thirty-year period? Does it exhibit some predictability forced by stable patterns? Or does it simply follow the dictate of profitable opportunities?

These questions were applied to international strategies implemented by the so-called "independent" oil companies. The international independent oil companies form a group that, surprisingly, has been almost completely ignored in the numerous studies concerning oil activities. Nevertheless, the growing importance of the independents occasionally has been emphasized. The point has been made that the concentration index within the international oil industry, whether in production, refining, or marketing, has decreased markedly with the independents entering successively in all phases. This change is shown dramatically in Exhibits 1 and 2.

This study is an examination of the independents' international strategies from two points of view: (1) the initial decision to go abroad; (2) the subsequent decisions dealing with foreign investment or divestment. Although these two research questions will be considered separately, they are chronologically and conceptually linked. Whereas the objective of the first research question is to explain one punctual move, that is, the point A on the time vector pictured in Exhibit 3, the second research question considers the whole continuum of strategic decisions from A to B.

II. The Sample Definition

The subjects of this study raise two problems of definition: one of sample size, which will be solved later in this section; the other one of validity of use of the label "independent," which will be dealt with first.

'Independent' has been used variously as a descriptive term for an operating company or individual. Prior to 1911, an independent was one who did not belong to the Standard Oil Trust or to any great foreign corporation. Several meanings have been attached to it since. In the United States, an independent is defined as one individual who "is not associated with or dependent

Exhibit 1

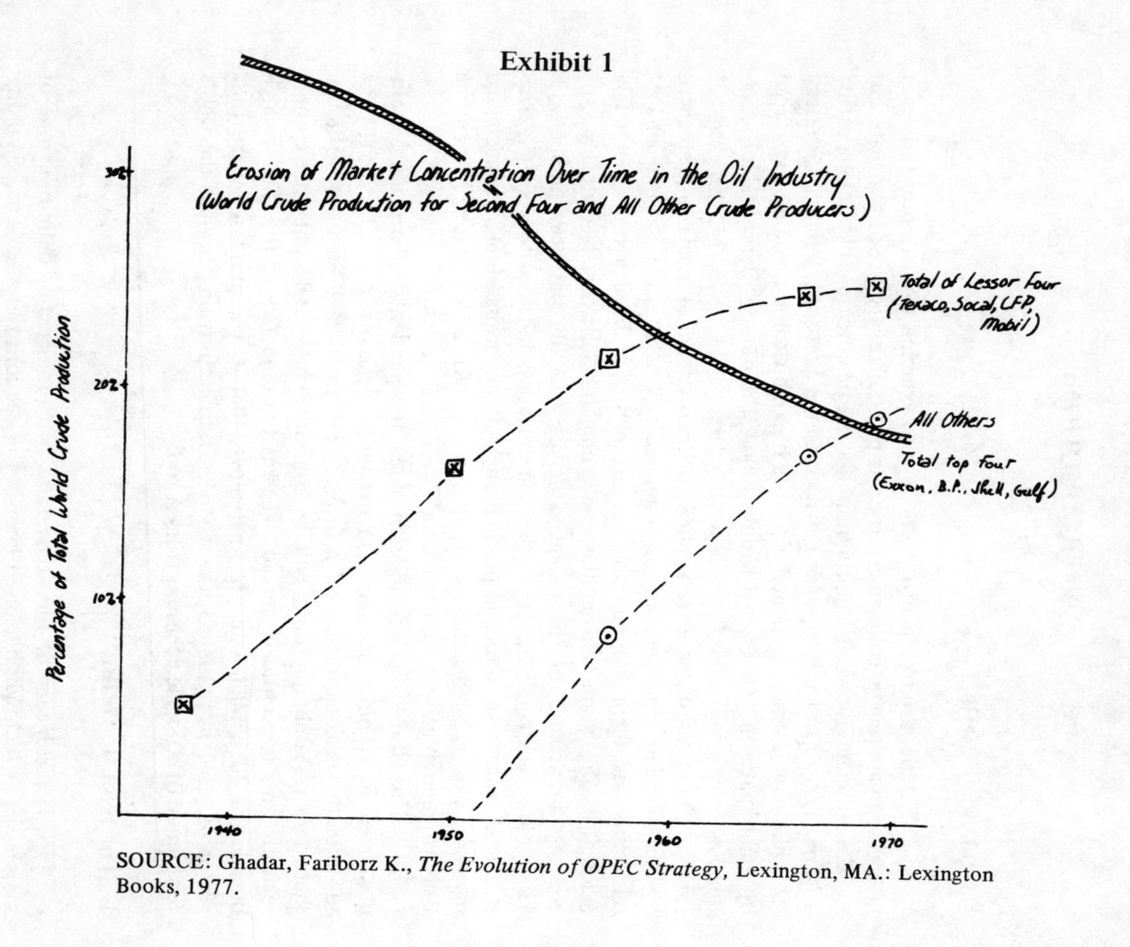

SOURCE: Ghadar, Fariborz K., *The Evolution of OPEC Strategy*, Lexington, MA.: Lexington Books, 1977.

Exhibit 2
World Refinery Capacity for Leading Four, Second Four, All Others Combined

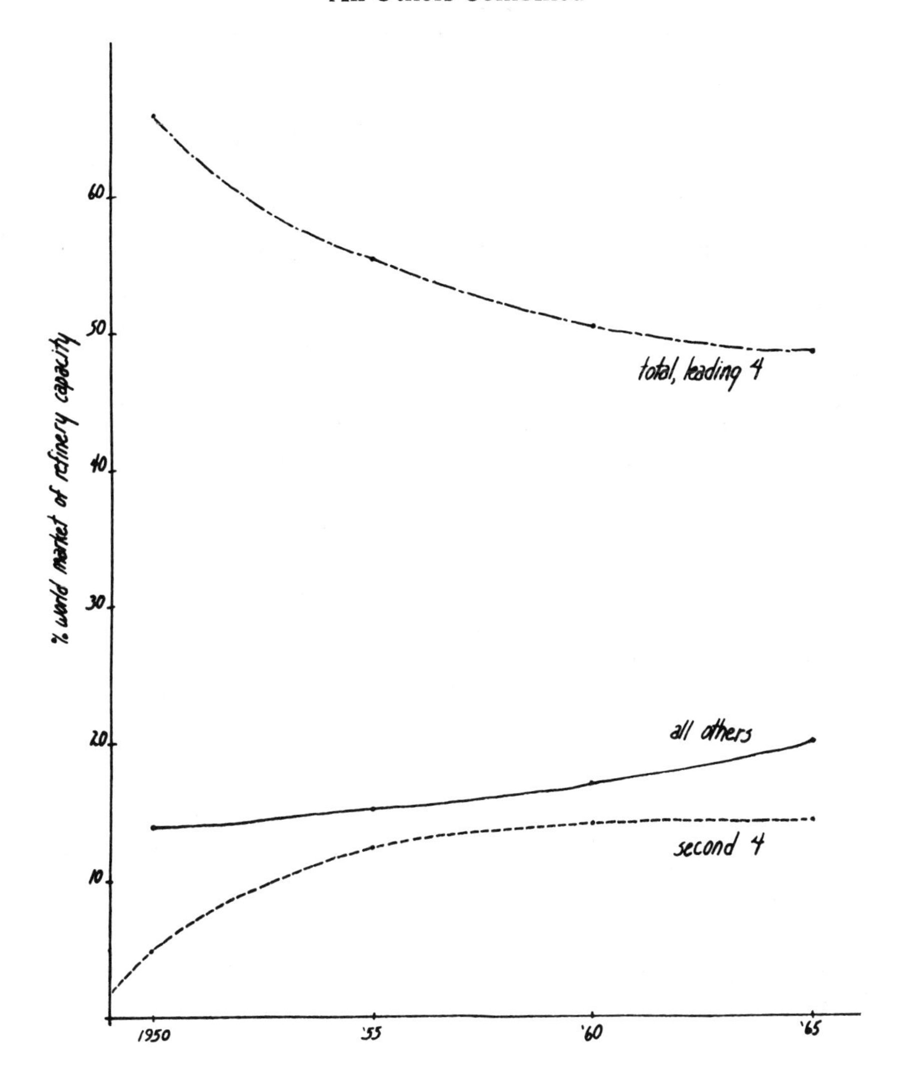

SOURCE: Ghadar, F., op. cit.

Exhibit 3–The Time Framework Underlying The Research

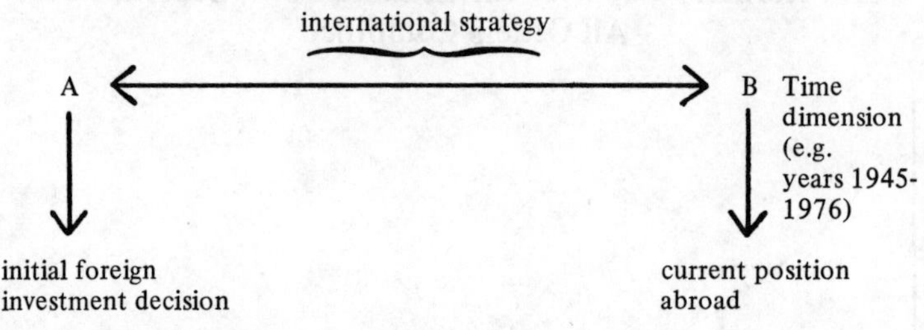

upon an integrated company."[1] In some cases it is any oil company or individual operator other than a major.[2]

In order to make a distinction between 'domestic' and 'international' independents, some scholars have rejected the label 'independent' in favor of others: 'smaller oil companies' (Vernon & Wells, 1976; Wilkins, 1975), 'international minors' (Rothberg, 1974; Sturgeon, 1974), or even 'the others' (Jacoby, 1974). According to Sampson, who called the majors the 'seven sisters,' 'independent' is a confusing name, but he preferred 'intruder,' which is no less disputable.

The search for an operational definition[3] can be solved in two ways. First, Vernon has suggested that, "as a rule, the term is applied to any large oil company standing outside the group that was operating in the Middle East at the end of World War II."[4] Indeed, only the seven or eight majors were in that group. Second, more generally, the term 'independents' refers to oil corporations that (1) operate internationally and (2) are not subsidiaries of the eight major oil companies.

However, none of these definitions is sufficient, for none provides an upper limit to the number of companies involved. *Oil and Gas Journal* identified some sixty U.S. firms seriously involved in foreign wildcatting operations in 1955.[5] According to Jacoby, 350 different firms (U.S. and foreign) entered the international oil industry between 1953 and 1972.[6]

Because I wanted to study United States firms, this research is concerned solely with U.S. oil companies. After deciding to concentrate on publicly-owned corporations–which excludes privately-owned Hunt Oil, for example–two pairs of criteria have been used to select the companies to be surveyed.

First, I took the largest oil companies after the U.S. majors as listed by *National Petroleum News,* "Fact Book 1976" (U.S. Source) and by *Petroleum Economist* (British source: March 1976). Then I set a minimum foreign production of 10,000 barrels per day of crude oil in 1975. Firms like Kerr-McGee, Tenneco, and Texas Gulf were thus excluded from the sample.

Shell U.S. was also rejected because it is owned at 69% by Royal Dutch-Shell. But Standard Oil of Ohio (Sohio) was kept even though British

Petroleum bought a controlling interest in it in 1970, for prior to that year Sohio was an independent company. Although Signal sold all of its oil business in 1974, it has been kept in the sample also, for it was significantly involved abroad before the sale.

The second pair of criteria consisted of adding to the sample independents that have merged with some of the above companies. The rationale was mainly to increase the degree of freedom of the sample. These additions were done, however, with the provision that the merged companies had had exploration concessions in at least four foreign countries in the 1950s.

The twenty-two independents thus selected are listed in Exhibit 4.

Exhibit 4–List of the Independents in the Research Sample

Amerada-Hess
Ashland
Atlantic-Richfield
Cities Service
Continental
Getty
Marathon
Murphy
Occidental
Phillips
Pure Oil (merged into Union Oil of California in 1965)
Richfield (merged with Atlantic in 1966)
Signal Companies
Sinclair (acquired by Atlantic-Richfield in 1969)
Skelly
Standard Oil of Indiana
Standard Oil of Ohio
Sun Oil
Sunray (merged with Sun Oil in 1968)
Superior Oil
Tide Water (merged into Getty in 1967)
Union Oil of California

It could be argued that, depending on the definition, these twenty-two companies do not represent an adequate sample. Specifically, in reference to the first research question, the test of hypotheses for going abroad should include also those companies that did not go abroad. Therefore restraining the study to those that went abroad on the basis that I am examining only international firms is tautological.

However, the initial investment decision (first research question) is considered here merely as the first of a chain of decisions (second research question). In other words, the standpoint of this study is not: "why going abroad?" in contrast to "why not going abroad?" but "why taking this first step of an international strategic continuum?" It was therefore justified to consider only those companies with a record of international strategy over a long period of time.

The sample selection process can be pictured in the following way:

- considering the worldwide oil industry as a square;
- the international oil industry–that is, each company's foreign activity–as a circle inside the square;
- the left part of the square as the U.S. oil industry;
- the right part as the foreign industries;
- the upper part as the majors;
- and lower part as the independents, the sample of the worldwide oil industry studied here is represented by the shadowed area in Exhibit 5.

III. Overview on Methodology

During the course of this research, two methodologies were used, one corresponding to each of the two research questions. To deal with the independents' first entry abroad, hypothesis testing has been used. To sort out patterns of international strategy over time, an exploratory approach followed eventually by hypothesis testing has been chosen. This dual choice was dictated by the nature of the two research questions. Whereas the first issue rests on a large body of knowledge, this is less true for the second topic.

The exploration was conducted in two steps. The first step was to compile a record of the investment decisions abroad that form the independents' international strategies. No such records existed before on this group of firms. The second step was to sort out patterns among international strategies into strategic groups, that is, sub-samples made of companies that followed identical strategies. In other words, from a methodological standpoint, this analysis was first longitudinal (each company separately over the period), and then cross-sectional (comparison of all the companies).

In this process, the approach was historical; that is, the method of research was to put together data of time-series type, both qualitative and quantitative, and then to form hypotheses about what policies they express with the help of a conceptual framework that will be presented in Chapter 4.

The research methodology involved also: 1) a search of literature concerning–a) the foreign direct investment decision, b) international strategy, and c) strategy in general; 2) an historical characterization of the international petroleum industry from published sources.

For the first research question, the statistical data used come from the surveyed companies' annual reports, supplemented in one case by the U.S. Bureau of Mines for a statistic on U.S. crude oil imports. For the second question, the method used to gather data was a combination of library research and

Exhibit 5—Scope of the Study

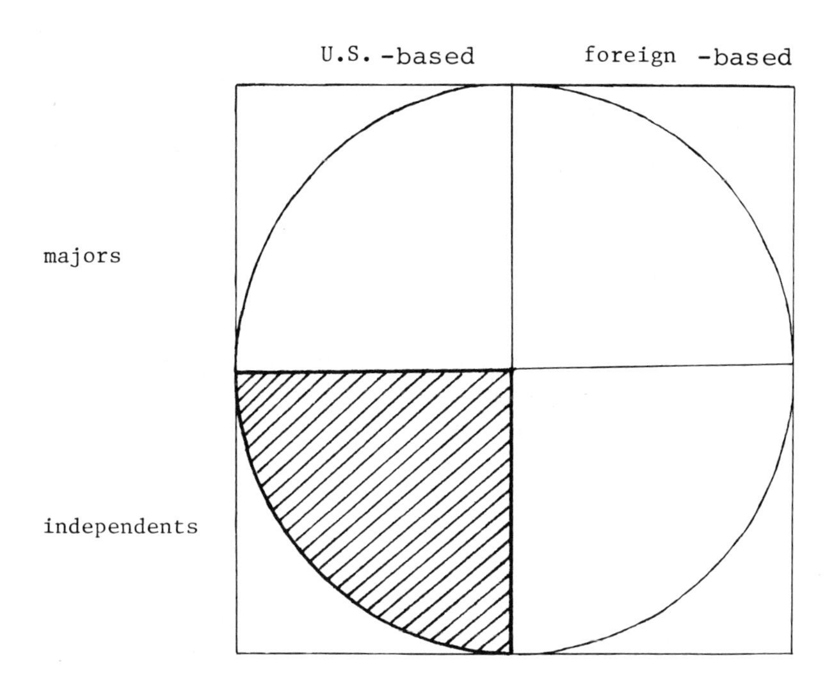

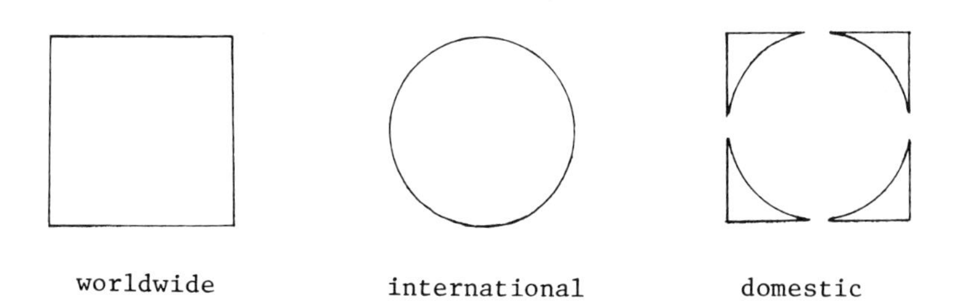

field interviews. The main source was the companies' annual reports, since the 1930s, which I studied systematically. I used *The Wall Street Journal, The New York Times,* and *Oil and Gas Journal* to cross-examine the annual reports' contents and supplement them when they failed to report key outcomes. Articles in other magazines, like *Fortune* and *Business Week,* were also used.

For contact with the companies, the interview process was selected rather than survey questionnaires. The exploratory nature of the second research question required that I obtain information on more policy details than would be possible with mailed questionnaires. The interview tactic chosen was to approach each company with a very detailed chronology of its moves abroad in the different phases of the oil business since the company's first entry abroad and to ask for rationale of these moves.

Fourteen companies from a total of seventeen existing independents were visited. For the five independents–Pure Oil, Richfield, Sinclair, Sunray, and Tide Water–that merged in the surviving independents, their strategies abroad were simply documented from published sources.

IV. Organization of the Manuscript

The study has been divided into two main parts, one for each research question. Part I deals with the initial investment decision.

Chapter 2 is an interpretive review of the theories on foreign direct investment and a suggested synthesis of the relevant literature on the international oil industry. Alternative or complementary approaches relevant for the independents' first entries abroad are described briefly with an indication of the potentials and drawbacks associated with each. The historical facts that emerge in the process help set the stage for a more quantitative look at the first entries abroad. A hypothesis is formulated at the end.

In Chapter 3 some key characteristics of the independents in their year of entry abroad are measured and tested. A hypothesis is then submitted to a multiple regression test. In the whole chapter the approach is quantitative.

Part II deals with the chain of subsequent foreign investment decisions forming the independents' international strategies.

Chapter 4 provides an interpretive review of theories about strategy in general and international strategy in particular as found in the disciplines of international business, industrial organization economics, and business policy. A model for the analysis of the independents' international strategies over a long period of time is then suggested.

Chapter 5 applies this model of "multinationalization through verticalization" to two periods: the thirty years before World War II and the thirty years after. Patterns in both periods are characterized and compared.

Chapters 6 and 7 contain an examination of the twenty-two independents with regard to their international strategies between 1945 and 1976. The major purpose is to look for repetition of patterns. Accordingly, the analysis resulted in two main strategic groups, which explains the allocation of the sample into two chapters. Characteristics in each group are submitted to tests of significance.

Chapter 8 as a general conclusion gives a summary of the major findings of the study and an assessment of their implications for corporate managers. Finally, several research issues are identified and ranked in order of urgency as topics for further research.

NOTES

[1] U.S. Congress, House, Select Committee on Small Business. "Effects of Foreign Oil Import on Independent Domestic Producers," (81st Congress, 1st. session; Washington: U.S. Government Printing Office, 1949-1950).

[2] Thereon 'majors' are defined by the following list: Exxon, Gulf Oil, Mobil, Texaco, Standard Oil of California, British Petroleum, Royal Dutch-Shell; and Compagnie Française des Pétroles.

[3] "A definition is an operational definition to the extent that the definer (a) specifies the procedure (including materials used) for identifying or generating the definiendum, and (b) finds high reliability for [consistency in application of] his definition." Julian L. Simon, *Basic Research Methods in Social Science,* (New York: Random House, 1969).

[4] Vernon, Raymond, *The Oil Crisis,* (New York: Norton, 1976), p. 4.

[5] Duff, Dahl M. "More Independents are Going Abroad," *Oil and Gas Journal,* vol. 53 (May 30, 1955); 46.

[6] Jacoby, Neil H., *Multinational Oil: A Study in Industrial Dynamics,* (New York: MacMillan, 1974), p. 126.

Part I

The First Decision to Invest Abroad

CHAPTER II

THEORIES AND FACTS ON THE INDEPENDENTS' FIRST ENTRIES ABROAD AFTER WORLD WAR II

Introduction

This chapter deals with both the theoretical and factual backgrounds surrounding the U.S. independents' entries abroad after World War II. I do not focus at this stage on the experience abroad of some independents before World War II, for the emphasis is on the postwar conditions. This factor is, however, incorporated in the hypothesis that is tested in the next chapter, and the issue is dealt with in more detail later.

In the first section of this chapter, I check what contribution the existing literature on foreign direct investment makes toward an explanation of the phenomenon studied. I review briefly the explanations suggested in the literature dealing with the oil industry, and especially its international component, in a second section. In a third section, I provide pro and con arguments about each factor that may have influenced the companies' decision, in an effort to put the reader in the atmosphere of the late 1940's economic constraints. In a fourth section, I acknowledge the justifications offered by the independents to their shareholders for their geographical diversification. The fifth section is devoted to the formulation of a hypothesis concerning the internationalization of the independents.

I. The International Business Literature on Foreign Direct Investment

In this section I chose not to present a detailed survey of the literature, for exhaustive ones are readily available.[1] The effort here was rather directed toward theories of foreign direct investment that could be relevant to the initial decision to go abroad after World War II.

Surprisingly enough, this search was inconclusive. Except for the "follow-the-leader" hypothesis, none of the available foreign direct investment theories provide alternatives to the hypotheses spotted in the petroleum literature dealing with the independents.

While the literature on foreign direct investment is abundant, there is as yet no generally accepted theory to explain why foreign direct investment takes place. Vernon has warned the potential researcher: "In an area as complex and 'imperfect' as international trade and investment, one ought not anticipate that any hypothesis will have more than a limited explanatory power."[2]

One approach to explaining first entries abroad has been to ask the companies themselves to identify their motivations. From such questionnaire studies made in the early sixties, it appears that the most frequently cited reasons for setting up foreign manufacturing operations were: (1) better market prospects, and (2) higher profitability.[3] These answers fall into a division that is not fortuitous. It corresponds to the two underlying competitive theories about the firm's objective function—(1) growth maximization, and (2) profit maximization—that will appear later on in this chapter behind some of the factors that drove the independent oil companies abroad.

The search for more profitable uses of capital is central in the portfolio theory of foreign direct investment.[4] This model contends that the share of its assets a firm will hold abroad depends on the relative rates of return at home and in foreign countries. Although intellectually appealing, the model by itself offers no theory as to why foreign direct investments become more or less profitable relative to domestic investments. Taking a risk vs. return factor into account, the portfolio theory looks promising because the oil business is a risky one. Such a hypothesis, nevertheless, would not explain why a group of firms in the same industry—the risk/return factor being identical for all the members—went into the same foreign countries at the same time, as the U.S. independents did.

The product life-cycle approach to foreign direct investment is concerned with manufacturing firms that start with direct export and then become multinationals through direct production abroad. Assuming that the independents went abroad into marketing before getting into production, the product life-cycle concept could be useful in accounting for some of the decisions the independents made. The product life-cycle model suggests that, in the last stage, the less-developed countries—with low production costs—will export toward the advanced countries. More specifically, Stobaugh hypothesized that "the nations having low-cost raw materials because of large oil and gas supplies might become major exporters of mature petrochemicals."[5] To be sure, this hypothesis would still be valid for oil companies in less developed countries—working either under concession terms or under service contracts.

The oligopoly approach is more fruitful in that respect. Hymer pioneered the concept.[6] He suggested that the motivating force behind foreign direct investment is that, by acquiring control over foreign enterprises, oligopolistic firms can reduce competition and increase profits. In the same line of thought, Vernon showed that the bulk of FDI was made by large firms in concentrated industries.[7] Testing Hymer's hypothesis for the first time, Knickerbocker found evidence that entries by U.S. firms into foreign markets were bunched in time—more so than could be expected by chance.[8]

Knickerbocker, however, was quick to point out the limit of the explanatory power of oligopolistic reaction: "If firm A makes a move, say investing abroad, then the concept of oligopolistic reaction tells something about what firm B, a rival of A, is likely to do. But it does not explain why firm A moved in the first place."[9] In other words, follow-the-leader implies by definition a leader A who would have invested abroad for a different reason than B. But it may also well be that B invested abroad for the same reason—a common threat or a common incentive—that A did in the first place. That does not mean, however, that the two explanations are exclusive: The fact that B responded to the threat/incentive after A did can be taken as a clue for qualifying the way B reacted compared to A.

II. Literature on the Independent Oil Companies

Jacoby summarized with a critical eye the existing literature on petroleum: "Although the foreign petroleum industry is the subject of an immense literature, most of it is topical and journalistic in character. That part which has scholarly value consists, in the main, of studies of particular divisions or aspects of the world, or it addresses the history, technology, or politics of the world oil industry."[10]

After an extensive reading of the literature (see Bibliography in Appendix at the end of the thesis), I tend to agree with this opinion. Indeed, many scholars have analyzed the "majors." They have acknowledged, incidentally, the existence of the independents, but few have really systematically examined this group of firms. It is notable that in the 1950s only one book dealt—though marginally—with the international independents; in the 1960s, there were five; and in the 1970s, seventeen—twelve being published after the oil crisis of 1973. The first book devoted mainly to the independents is dated 1974.[11] Yet almost all suggest different hypotheses—although not tested—for the entry of the independents on the international scene after World War II.

In order to avoid a tedious survey of literature, I simply presented the relevant contributions in a synthesized form in Exhibit 1. In all, scholars and business journalists have suggested ten distinct factors in their effort to explain the independents' entries on the international petroleum scene. They are summarized in Exhibit 2.

III. The Conditions Surrounding the Independents' Initial Entries Abroad

The purpose of this section is to present pro and con arguments about each factor of entry listed in Exhibit 2. None of these explanations have been tested yet, but not all need to be, for some have actually a weak explanatory power, as I will suggest.

Exhibit 1
Synthesis Survey of Relevant Literature on Independent Oil Companies

Date of Publication	Author(s)	Title of the Book	Perspective and Methodology	Hypothesis (explicit; E) (implicit; I)
1959	DECHAZEAU & KAHN	Integration and Competition in the petroleum industry	business economics/ industrial organization	foreign tax credit (E)
1965	BARROWS	International Petroleum	descriptive-historical	rising costs of finding oil at home (E)
1966	FRANKEL	Mattei, Oil and Power Politics	historico-journalistic	personality of the company's top manager (I)
1966	DONALDSON et al.	The International Oil Industry	financial analysis	higher return abroad and increased demand in Euope (E)
1967	GABRIEL	The Gains to Local Economy(): Case of Venezuela	business economics	higher return abroad and decreased barriers to entry abroad (E)
1968	PENROSE	The International Petroleum Industry	business economics	host government oil policy (I)
1971	VERNON	Sovereignty at Bay	business economics	host government oil policy (E) and decreased barriers to entry (E)
1972	ADELMAN	The World Petroleum Market	business economics	(provides no explanation)
1973	SHWADRAN	The Middle East, Oil and the Great Powers	descriptive-historical	(provides no explanation)
1973	MEDVIN	The American Oil Industry	business economics	higher return abroad (I)
1973	MOSLEY	Power Play–Oil in the Middle East	historico-journalistic	personality of company's top manager (I)
1974	CHURCH Committee	Congress Hearings	historico-episodical	(provides no explanation)
1974	JACOBY	Multinational Oil	industrial organization	higher profit abroad and growth of the market and company's own oil reserves maximization and host government oil policy and decreased barriers to entry and depletion incentive (all E)

Exhibit 1

Synthesis Survey of Relevant Literature on Independent Oil Companies (continued)

Date of Publication	Author(s)	Title of the Book	Perspective and Methodology	Hypothesis (explicit; E) (implicit; I)
1975	DUCHESNEAU	Competition in the U.S. Energy Industry	industrial organization	foreign tax credit (E)
1975	KRUEGER	The U.S. and International Oil	descriptive-historical	(provides no explanation)
1975	RAND	Making Democracy Safe for Oil	political-historical	(provides no explanation)
1975	F.E.A.	Relationship of Oil Companies and Foreign Governments	historical-survey	(provides no explanation)
1975	VALLENILLA	Venezuelan Oil and OPEC	historico-political	increased world demand (E)
1975	SAMPSON	The Seven Sisters	historico-journalistic	personality of company's top manager (I)
1975	U.S. CONGRESS	Background Readings on Energy Policy	survey	foreign tax and royalty credit (I)
1976	SOLDBERG	Oil Power	economico-political	rising costs of finding oil in the U.S. (E); tax credit for royalties (E) increased demand for oil in Europe/ Japan (E)
1976	VERNON (ed.)	The Oil Crisis	various	threat of imports of cheap oil by competitors (E)
Sources which do not talk about Independents but suggest trends in environment				
1949	U.S. CONGRESS	Effects of Foreign Oil Imports on Independent Domestic Producers	survey-hearing of testimonies	cheap cost of import
1949	MIKESELL & CHENERY	Arabian Oil–America's Stake in the Middle East	business economics	(mentions non-U.S. majors, but only in South America) lower cost of getting reserves abroad
1952	U.S. CONGRESS	The International Petroleum Cartel	survey-historical	potential of Middle East
1954	MCLEAN & HAIG	The Growth of Integrated Oil Companies	industrial organization	(deal only with the domestic operations of some independents)

Exhibit 2
EXPLANATIONS SUGGESTED FOR THE INDEPENDENTS' MOVES ABROAD

Factors of Entry Abroad*	Analytical Perspective
1. Tax incentives (deChazeau, Duchesneau, Jacoby, Soldberg, U.S. Congress)	environment
2. Personality of company's top executive (Frankel, Mosley, Sampson)	organizational behavior
3. Host government's oil policy (Jacoby, Penrose, Vernon)	environment
4. Decreased barriers to entry (Gabriel, Jacoby, Vernon)	industrial organization
5. Increased demand for oil (Donaldson, Jacoby, Soldberg, Vallenilla)	environment
6. Potential for oil areas abroad (Jacoby)	environment industrial org.
7. Higher return abroad (Donaldson, Gabriel, Jacoby, Medvin)	industrial organization
8. Increasing cost of finding oil in the U.S. (Soldberg)	environment industrial org.
9. Company's oil reserves expansion (Jacoby)	strategy
10. Threat by competitors (Vernon)	industrial organization

*The names in parentheses refer to the references listed in Exhibit 1. All the above possible factors of entry are, however, described in detail in the following section, for they provide a historical background necessary to understand the conditions surrounding the independents' entries abroad in the late forties and early fifties.

Factor # 1–Tax Incentives. The origin of the tax incentives has been thoroughly described. When, in 1948, Getty's Pacific Western Oil Corporation negotiated for a one-half interest in the Neutral Zone, Getty agreed to pay Saudi Arabia a royalty of fifty-five cents per barrel, whereas Aramco was paying only twenty-one cents. Saudi Arabia immediately demanded more money from Aramco's shareholders. Turning to the U.S. government for assistance, Aramco was advised that a tax solution existed. Since the Revenue Act of 1918, the United States has allowed a foreign tax credit against income derived from foreign sources. In 1950 the Internal Revenue Bureau ruled that a 50% tax on petroleum companies, levied by Venezuela, could be treated in this manner. Following this example, the Saudi Arabian government imposed a 50% income tax on Aramco late in 1950. Similar arrangements were incorporated in the Iranian Oil Consortium agreement in September 1954, and the Internal Revenue Service approved the Aramco scheme in May 1955.

Soldberg contended that because of this tax incentive, "majors and independents alike turned their backs on domestic exploration after 1950 to drill abroad."[12] If it were so, it would show that the tax incentive was not a determinant factor in the independents' moving abroad. Firstly, it did not apply specifically to them because they had no production abroad but, instead, to the U.S. majors with existing foreign production, who asked for it. Secondly, the Church committee pointed out that the first tax credit treatment was claimed by oil companies in 1950, but the first published ruling by the Treasury waited until 1955; so there was a five-year lag before the generality of the tax deduction was a published fact. Even assuming that, by 1951, the whole petroleum industry was aware of the new tax development, it still does not apply to the independents who went abroad in majority before 1951 and, furthermore, into exploration and not directly into production.

Factor #2–Personality of Company's Top Executive. While two authors (Frankel and Mosley) subscribed implicitly to the personality hypothesis, only Sampson made it explicit: "The significance of Getty is that he was the first individual to challenge the dominance of the Seven Sisters in the Middle East, thus helping to begin the erosion of their monopoly. He typified the spirit of the rising independents."[13] Trying, however, to explain the independents' going abroad by the aggressive behavior of their top executives, while appealing, is limited. Paul Getty and Armand Hammer may have been decisive factors in their companies' going abroad, but the former was a leader among the independents while the latter was a follower. Yet the personality hypothesis is worth consideration as a descriptive item in a historical perspective. In this research, however, the personality factor will not be dealt with.

Factor # 3–Host Government's Oil Policy. Drastic changes occurred in the producing countries' oil policies after World War II. Before, concession agreements were especially negotiated, covered large areas and very often the entire country, were of very long duration (up to ninety-nine years), and resulted in one or a few oil companies–generally the majors–being given control over a country's entire oil resources. After World War II, petroleum laws were passed by most countries to standardize the terms of all exploration and development concessions. In particular, areas conceded to any one firm were sharply limited in space and leaseholds reduced in time. By setting such limits, the producing countries could attract more petroleum companies. In the fifties, two countries–Libya and Venezuela–were in the forefront of this policy change, and the independents took advantage of the opportunity. But it was mere additional incentive to go on exploring for, as mentioned already, the independents had made their first move abroad much earlier.

If the countries' invitations were a necessary condition for the independents' entry, it was not, however, automatically a sufficient one, for the companies could elect not to accept the invitation.

Factor # 4–Decreased Barriers to Entry. These changes in host countries' oil policies lowered some of the barriers to entry abroad. But other barriers were believed to remain for a long time. According to Gabriel, "development of the deposits required such large investment, . . . involved such large initial risks, and took place under concessions of such magnitude, that the large firms enjoyed an overwhelming advantage in the financial, technical and risk-reducing economies of scale associated with size. . . . Recently the cost of securing a foothold in foreign producing has declined, so that the barriers to entry into international producing have come down. Also an increasing number of firms have reached the requisite size for participation."[14] Vernon pointed out that the emergence of national oil companies was "tangible evidence of the fact that some of the long-time barriers to entry into the industry at the exploitation and refining stage had at least been lowered."[15] Also, as early as 1949, Mikesell and Chenery observed that "in foreign production, then, financial, technical and risk-reducing economies all favor large firms."[16] Yet, lumpiness of the foreign investment was not necessarily an absolute deterrent for the small firms, for they could share the financial burden through joint ventures with other small firms. Indeed, as shown in Exhibit 3 based on data from 1952, the majority of the independents studied represented less than one-fifth of the average U.S. major in terms of total assets or income, and less than one-third in terms of domestic production or refinery runs.

Factor # 5–Increased Demand for Oil. According to Vallenilla, the independents "became interested in international operations because of increased world demand for oil during the postwar era, offering a possibility to increase sales within the U.S. and at the same time capture new markets abroad."[17]

The growth of demand for petroleum products is a natural explanation for the efforts of the suppliers. If the prospect of a rising demand was a factor, however, it was only the case of the domestic demand, for when they went exploring abroad, the independents did not intend to serve anything but the U.S. market, nor did they have the facilities to do so.

Factor # 6–Potential of Oil Areas Abroad. At the end of World War II, there was a movement toward presenting foreign oil areas as opportunities, as witness titles of several articles published in *Oil and Gas Journal*: "Design for petroleum expansion outside the U.S."[18]; "Petroleum-supply problem is world wide"[19]; "Oil No. 1 opportunity in South America"[20]; "Vast areas in Middle East awaiting exploration and development"[21]; "World development points further toward one-world conception."[22]

In 1946, Joseph Pogue, Vice President of the Chase Manhattan Bank, estimated that future Middle East discoveries would reach 150 billion barrels compared with 27 billion barrels of estimated proved reserves. For the U.S. the figures were respectively 50 and 21 billion barrels.[23] E. DeGolyer, an internationally known geologist who had led Amerada in the 1920s, confirmed that

Exhibit 3

SIZE OF INDEPENDENTS IN 1952

	Total Assets		Total income		Net income		Net domestic production		Domestic refinery runs	
	(1)	(2)	(1)	(2)	(1)	(2)	(1)	(2)	(1)	(2)
St. Indiana	39.3	83.5	38.9	80.9	23.1	50.5	60.5	87.1	65.0	96.8
Cities Service	21.6	45.9	21.9	45.4	9.5	20.7	20.7	29.9	28.3	42.4
Sinclair	20.5	43.6	21.0	43.7	16.6	36.4	27.1	39.0	46.4	69.1
Phillips	18.3	38.8	27.4	36.2	14.5	31.7	30.5	43.9	28.0	41.7
Atlantic	10.4	22.1	14.8	30.7	7.8	17.0	22.4	32.2	22.7	33.8
Union	8.7	18.5	6.9	14.4	5.3	11.6	22.6	32.5	18.3	27.2
Sun	8.6	18.2	14.9	31.0	8.3	18.1	26.0	37.5	27.4	40.8
Pure Oil	7.6	16.0	8.2	17.1	5.3	11.5	17.9	25.7	14.5	21.6
Continental	7.1	15.2	2.6	20.0	7.3	16.0	28.0	40.3	14.2	21.1
Tide Water	6.8	14.4	10.4	21.6	6.0	13.1	23.6	33.9	19.3	28.7
Sohio	5.9	12.5	7.3	15.2	3.4	7.4	7.9	11.4	15.2	22.7
Marathon	5.6	12.0	5.4	11.3	7.6	16.6	22.7	32.7	4.8	7.2
Skelly	5.0	10.6	5.0	10.4	5.4	11.8	16.0	23.1	5.5	8.3
Richfield	4.7	10.1	4.4	9.1	4.9	10.8	14.3	20.5	13.5	20.1
Amerada	1.8	3.9	1.9	3.9	3.1	6.7	16.9	24.3	–	–
Average 'Independent'	11.5	24.4	12.5	26.1	8.5	18.7	23.8	34.3	23.1	34.4

(1) Relative size in terms of Exxon's.
(2) Relative size in terms of the five U.S. majors.
SOURCE: The figures are derived from data published in John C. McLean and Robert W. Haig, *The Growth of Integrated Oil Companies* (Boston: Division of Research, Graduate School of Business Administration, Harvard University, 1954), p. 332. Data were not available in this source for Ashland, Getty, Murphy, Signal, Sunray, and Superior.
All data are in percentage.

the Middle East actually had been "very little explored."[24] Julius Fohs, another well-known geologist, added that, beyond the primary exploitable oil reserves held mainly by the majors, there were sizable second- and third-class oil areas, making "a total of 975,000 sq. miles of exploitable territory for oil in the Middle East."[25]

Factor # 7–Higher Return Abroad. According to Jacoby, "The relatively high rates of return earned by the largest foreign oil companies up to 1955 constituted strong inducements to new firms to enter the industry."[26] Indeed, profit computed on the book-value of U.S. direct investment abroad showed that for all petroleum operations the return was 18.5% worldwide in 1950. At the production stage, it was 25.6% in Venezuela and 32.4% in the Middle East.[27]

Assuming that managers of domestic companies were aware of the profit discrepancy in favor of foreign involvement, it must have been a factor.

In any case, such higher profitability could be expected. As one Aramco executive put it before the Congress in 1949: "You have to realize that an enormous amount of money has to be put into an enterprise of that kind, much of it put up initially by gambling companies."[28] Indeed, higher returns abroad exist because of higher hurdle rates, which mean higher expected risks, especially in exploration. Furthermore, the time lag between the first exploration investment abroad and the first discovery is known to be long. For these reasons, higher risk goes with higher reward, and I will assume that such was the general expectation of the independents in venturing abroad.

Factor # 8–Increased Cost of Finding Oil in the U.S. Compared with Abroad. Back in 1949, Mikesell and Chenery were already pointing out this phenomenon: "Although proved reserves in the U.S. are at an all-time high, their ratio to annual production has been declining since 1938. . . . The number of exploratory wells necessary to find adequate new reserves has constantly increased. Between 1936 and 1946 the oil discovered per wildcat well drilled declined from 1,180,000 barrels to 160,000."[29]

The U.S. productivity per well looked even worse when compared with the figures obtained abroad. The average production per well per day in the late 1940s was: 12.7 b/d in the U.S.; 214.4 in Venezuela; 32.3 in other countries of the Western Hemisphere; 4,232.9 in the Middle East; 65.8 in other countries of the Eastern Hemisphere.[30] As one witness put it before the Congress: "A hole in the ground costs about the same whether it be in Arabia or in Texas. If he gets 6,000 b/d out of one hole and 17 b/d out of the other, the fellow producing the 6,000 b/d can whip you to death on your 17 b/d."[31] To go abroad was therefore an obvious solution, and even more so since "in 1945 the capital expenditure required abroad was only 43 cents per barrel compared to 78 cents at home,"[32] according to one oil expert's estimations. Indeed, it is the same to say that increased costs at home drove the independents abroad as to say that higher profits abroad attracted them. Although the distinction was made by some scholars, I will consider the eighth point as a subpoint of the seventh.

Factor # 9–Company's Oil Reserves Expansion. Jacoby is the only author to have mentioned this explanation, although the interviews I made showed clearly that oil reserves expansion was a general and permanent concern among the independents: "The pressures of . . . competition tend to drive . . . each oil company to try to expand its own reserves, even though reserves in the aggregate appear to be ample. . . ." Companies with ample reserves in current producing areas nevertheless explore new areas in order to diversify their sources of crude oil for political, locational, or quality advantages. They seek to protect themselves against the possibility that rivals may discover large reserves of low-cost crude oil."[33] In particular, the facts exposed in the following paragraphs give high credit to this concern about the rivals' moves, primarily the majors'.

Factor # 10–Threat by the Majors. According to Vernon, in the early 1950s "most oil companies were content to limit their business activities to the U.S. But anxiety obliged some of them to begin looking around–anxiety stemming from the fact that some of the oil companies already established on a multinational basis had begun bringing some [cheap] oil into the U.S. from the Middle East. Faced with the competitive threat of a cheaper source of oil, some of the so-called independents responded in the standard way; their anxiety led them to take the plunge into distant foreign environments."[34] Since this explanation is incorporated in a more comprehensive hypothesis that is tested in the next chapter, I will just present here some historical facts that have been overlooked in the previous descriptions of the independents' entries abroad.

Even before the end of the War, the possibilities of crude imports by the majors existed, but had not materialized. As soon as hostilities ceased, there began a "considerable speculation as to whether Middle Eastern crude will, at some future date, move into markets on the U.S. East Coast."[35] For the purely domestic companies this prospect was not reassuring: "It is stressed that the economic factors involved in the Middle East crude movements, in relatively large amounts, always constitute a threat."[36] Indeed, "excessive imports could adversely affect the search for oil at home," a Vice President of Standard (New Jersey) acknowledged as early as 1945.[37]

While the majors kept repeating that they had no intention to harm domestic producers with their Middle East operations, they were moving in the expected direction. "Recently," a spokesman reported early in 1947, "it was made known that Texaco was influenced to locate its new refinery in the Philadelphia area because of the possibility that substantial imports of oil (from the Middle East through the Trans-Arabian pipe line) into the U.S. may be used."[38] The first tanker from Kuwait (115,000 b.) shipped by Gulf Oil to its Philadelphia refinery arrived on August 18, 1946: "To be used as a test run." In October 1947, Socony-Vacuum brought in the first of three tankers allegedly for the same purpose. But the experts of the time saw there the start of a new trend: "While nothing is being said officially about Middle East oil imports coming to the U.S. in increasingly substantial amounts, there is already evidence that at least one company, Socony-Vacuum, has passed the 'testing stage.' "[39] Indeed, Socony-Vacuum justified its imports as being done "to assist in alleviating the impending shortage of fuel oil and gasoline."[40] Standard (New Jersey) planned to follow, but could not immediately because of a lack of tankers. By May 1948, Socony-Vacuum had received its millionth barrel of crude oil from the Persian Gulf.[41] The first "experimental" shipment of Middle East crude to the West Coast was made by Standard Oil of California in January 1949. The same year, during congressional hearings on the effects of foreign oil imports on independent domestic producers, Standard Indiana, Cities Service, Sinclair and Atlantic–although not Middle East producers–were cited in a list of ten importing companies.[42] Tide Water was also mentioned as an importer back in the middle of 1948.[43]

Perhaps eager to scale down the importance of Middle East imports as threats for domestic oil companies, the majors tended to be optimistic–at least in the short-run–in their statistical forecasts. In July 1948, a Vice-President of Gulf Oil, for example, estimated "this year imports from Middle East to 30,000 barrels per day [actual figure: 63,700]; in 1949, around 60,000 b/d [101,900]; increasing to 150,000 b/d by 1951-52 [158,000 b/d in 1952]; and to 500,000 b/d by 1955 [269,800]."[44] The President of Standard New Jersey had concluded in 1948 that "investments for the purpose of permanently supplying the domestic market from the Middle East are not presently attractive, relative to other uses for this capital. Other investors, of course, may look at it differently. The important question is: Will capital be attracted to the Middle East to provide a permanent oil supply for the U.S.? We doubt it."[45]

But by early 1948 it was known that Middle East crude oil could be imported in the U.S. cheaply: "At present charter rates quoted by U.S. Maritime Commission, Saudi Arabian crude could be brought to East Coast ports at prices considerably under current domestic prices."[46] In terms of production costs, foreign crudes were more competitive than the U.S. In 1947 a report of Colombia's National Petroleum Council showed that the average Free-on-Board cost (in U.S. $) was: U.S., .645; Venezuela, .545; Colombia, .64; Ecuador, .53; Peru, .78; Iran, .27; Saudi Arabia, .315; Bahrein, .355; Iraq, .485; Kuwait, .28.[47]

Cost of transportation had to be added, however, which made the first shipments of Middle East crude justified only for test runs. For example, Socony-Vacuum revealed that these costs were "in excess of $2.50 per barrel," with transportation charges alone accounting for approximately $1.70.[48] Assuming completion of the Trans-Arabian pipeline by 1951, transportation cost should be decreased as shown in Exhibit 4.

Assuming use of a company's own tankers, both sources agreed that the maritime transportation cost would approximate 75 cents per barrel, bringing down the Arabian price to $2.065. The conclusion of this cost-accounting exercise is that: (1) not only could Middle East crude oil be brought into the U.S. East Coast more cheaply than Texas crude, but (2) Venezuelan crude, imported in the U.S. since the 1920s, was even cheaper.

Exhibit 4
Estimated Landed Price per Barrel on U.S. East Coast of Foreign and Domestic Crude Oil

Source of crude oil / Costs	Foreign Crude Oil: Saudi Arabia (1947)	Foreign Crude Oil: Saudi Arabia (after Trans-Arabian completion)	Foreign Crude Oil: Venezuela*	Domestic Crude Oil: West Texas crude (equivalent to Arabian landed costs in New York)
Actual production cost	$0.20	.20	.50	2.42
Royalty and taxes	.23	.23	.35	
Transportation	1.69	1.04	.20	.38
Pipeline cost		.30		.275
Loading and unloading charges	.03	.03		.03
U.S. import duty	.105	.105	.105	
Difference in refinery realization	.20*	.20*	.10	
Profit	.25*	.25*	.25	
	2.705	2.355	1.505	3.105

SOURCE: Estimation based on *Oil and Gas Journal*'s hypotheses (February 5, 1948, p. 46 and 49); "*" estimations based on Mikesell and Chenery's study (op. cit., 1949, p. 19).

IV. Companies' Justifications for Going Abroad

So far this chapter has dealt only with the explanations proposed by scholars and business journalists. But the companies studied also offered some reasons to their shareholders for their entry into the international scene. Fourteen of them justified their move abroad as shown in Exhibit 5.

In order to check for possible similarities or discrepancies between the scholars who observed the phenomenon and the independents who lived it, I compared the opinions of both parties as shown in Exhibit 6. It appears clearly that points of view are different. Whereas scholars emphasized environment aspects, companies accentuated management issues. But more important, the companies mentioned in majority[49] necessary and sufficient conditions of entry while the scholars named a majority[50] of secondary factors.

V. Formulation of a Hypothesis about the Independents' Initial Entries Abroad After World War II

The hypothesis, whose test is presented in the next chapter, is a combination of several factors, and it boils down to a profit-maximation and risk-minimization behavior. I would expect the managers of the independent U.S.

Exhibit 5
Independents' Rationales for Entry Abroad

	year of entry outside North America	
Amerada	1952	import threat (A/R 1954)[1]
Ashland	1948	"with increasing demands for petroleum in the U.S. () domestic crude supplies must continue to be supplemented by imports" (A/R 1947) / "development of crude reserves" (A/R 1951)
Atlantic	1945	"to augment further our controlled supplies of crude & reserves" (A/R 1943 & 1945)
Cities Service	1946	"effort to add to oil reserves" (A/R 1946 & 1954) potential of Middle East oil (A/R 1948)
Continental	1952	to "materially increase company's crude oil productive capacity" (A/R 1951 & 1956)
Getty	1949	(no reason given)
Marathon	1952	"rising cost of finding and developing new crude oil reserves in the U.S." (A/R 1955 & 1958)
Murphy	1956	(no reason given)
Phillips	1944	"acquiring supplementary low cost reserves of raw materials" (A/R 1945)
Pure Oil	1956	"possibility of greater returns for the effort" (A/R 1956)
Richfield	1946	"because proved reserves of recoverable oil in California are shrinking in relation to the demand, and in order to protect its business and its customers" (A/R 1954) import threat (A/R 1955)
Signal	1948	(no reason given)
Sinclair	1945	(no reason given)
Skelly	1958	(no reason given)
St. Indiana	1948	"If the increase in import continues it will be a further threat to the prosperity and growth of the domestic industry" (A/R 1949 & 1953 & 1956)
St. Ohio	1945	"the continuing rise in the cost of finding and developing crude oil supplies in the U.S." (A/R 1955)
Sun Oil	1956	"acceleration of foreign imports in the previous 6 months" (A/R 1952 & 1954)
Sunray	1957[2]	"search for new reserves; () potential of Canada and Middle East oil" (A/R 1948) import threat (A/R 1949 & 1954)
Superior	1947	(no reason given)
Tide Water	1956	"to find new sources of supply for meeting the U.S.' rapidly growing demand" (A/R 1955)
Union Oil	1947	(no reason given)

[1] A/R means "Annual Report." The companies' annual reports constitute the source for this exhibit.

[2] Although Sunray took an interest in Aminoil in 1947, it was reduced quickly afterwards so as to be considered from then on only a "minor stock interest" (A/R 1953). For this reason, Sunray's entry into Venezuela in 1957 was taken as first entry outside North America for the purpose of quantitative analysis. (See next chapter)

Exhibit 6
SCHOLARS' VS. COMPANIES' RATIONALES FOR ENTRIES ABROAD

Factor of Entry	Number of Times Mentioned By: Scholars	Companies
1. Tax incentives	5	–
2. Personality of company's top executive	3	–
3. Host government's oil policy	3	–
4. Decreased barriers to entry	3	–
5. Increased demand for oil	3	2
6. Potential of oil areas abroad	1	2
7. High return abroad	4	1
8. Increasing cost of finding oil in the U.S.	1	3
9. Company's oil reserves expansion	1	6
10. Threat by competitors	1	6

SOURCE: Exhibits 2 and 5.

oil companies to decide to undertake exploration abroad immediately after World War II because:

1) When the U.S. majors started to look for domestic outlets for the low-cost crude oil they produced in the Middle East, they threatened the independents' market positions in the U.S. The independents not only were relying only on more costly domestic crude oil, thus incurring the opportunity cost of not being a foreign producer, but in addition:
2) the increased cost of finding oil in the U.S. compared with abroad, measured by the ratio 'discovery plus productivity of wildcat well rates/exploration expenditures and expenses,' implied a profit penalty for a purely domestic firm. By going abroad, a firm will minimize the risk of not being successful enough in its domestic exploration as well as it will maximize its expected profits, given the higher productivity of wildcat wells abroad. The need for going abroad was comparatively more compelling for a crude short company concerned with profitability than for a crude long one. Indeed, assuming that a balance between production and refining is a permanent target, as studies on vertical integration have shown,[51] the crude short firm will have a more pressing need to find new sources of oil than the crude sufficient company, and thus the former will have a greater exposure to the

increased cost of domestic exploration. The unbalanced firm will have, therefore, a compelling reason to go exploring abroad in order to maximize its long-run profits jeopardized at home.

3) But, whatever its crude position in regard to its refinery needs, the bigger the independent (that is, the greater the quantities of oil produced), the greater will be the necessity to replace and expand its crude oil reserves–the constant search for additional reserves is a characteristic of the oil industry. At the same time, a big firm will be more able than a small one to skip the barrier represented by the lumpiness of the initial investment abroad, and therefore I would expect the biggest among the independents to go abroad first.

4) Foreign involvement prior to World War II would allow an independent to know by experience the advantages in terms of profit maximization and risk reduction of activity abroad, and thus I would expect such a company to re-enter the international petroleum scene more quickly than a company without prior foreign experience would make its first entry abroad.

NOTES

[1] See John H. Dunning, "The Determinants of International Production," *Oxford Economic Papers* 25 (November 1973); Gary Hufbauer, "The Multinational Corporation and Direct Investment," in *International Trade and Finance–Frontiers for Research,* ed. by Peter Kenen (New York: Cambridge University Press, 1975); Guy V.G. Stevens, "The Determinants of Investment," in *Economic Analysis and the Multinational Enterprise,* ed. by John H. Dunning (London: Allen and Unwin, 1974).

[2] Raymond Vernon, "International Investment and International Trade in the Product Cycle," *The Quarterly Journal of Economics* 80 (May 1966):198.

[3] Harry J. Robinson, *The Motivation and Flow of Private Foreign Investment* (Menlo Park, California: Stanford Research Institute, 1961); Raghbir S. Basi, *Determinants of U.S. Private Direct Investment in Foreign Countries* (Kent, Ohio: Kent State University, 1963).

[4] Martin F. Prachowny, "Direct Investment and the Balance of Payments of the U.S.: A Portfolio Approach," in *International Mobility and Movement of Capital,* ed. by Fritz Machlup (New York: National Bureau of Economic Research, 1972).

[5] Robert B. Stobaugh, "The Neotechnology Account of International Trade: The Case of Petrochemicals," *Journal of International Business Studies* 2 (Fall 1971):55.

[6] Stephen H. Hymer, "The International Operations of National Firms: A Study of Direct Foreign Investment" (Unpublished doctoral dissertation, M.I.T., 1960).

[7] Raymond Vernon, *Sovereignty at Bay: The Multinational Spread of U.S. Enterprises* (New York: Basic Books, 1971).

[8] Frederick T. Knickerbocker, *Oligopolistic Reaction and Multinational Enterprises* (Boston: Division of Research, Graduate School of Business Administration, Harvard University, 1972).

[9] Ibid., p. 8.

[10] Jacoby, op. cit., p. xxiv.

[11] Jacoby, op. cit.

[12] Carl Soldberg, *Oil Power* (New York: Mason/Charter, 1976), p. 9.

[13] Anthony Sampson, *The Seven Sisters* (New York: Viking Press, 1975), p. 141.

[14] Georg K. Gabriel, "The Gains to the Local Economy from the Foreign-Owned Primary Export Industry–The Case of Oil in Venezuela" (Unpublished doctoral dissertation, Graduate School of Business Administration, Harvard University, 1967), p. 61.

[15] Vernon, *Sovereignty at Bay,* p. 36.

[16] Raymond F. Mikesell and Hollis B. Chenery, *Arabian Oil, America's Stake in the Middle East* (Chapel Hill, North Carolina: University of North Carolina Press, 1949), p. 148.

[17] Luis Vallenilla, *Oil: The Making of a New Economic Order* (New York: McGraw-Hill, 1975), p. 82.

[18] *Oil and Gas Journal,* June 15, 1946, p. 84, by a representative of Union of California.

[19] Ibid., May 17, 1947, p. 62, by L. F. McCollum, to become top executive of Continental Oil.

[20] Ibid., May 31, 1947, p. 113.

[21] Ibid., July 19, 1947, p. 46.

[22] Ibid., December 13, 1947, p. 63.

[23] Ibid., August 8, 1946, p. 59.

[24] Ibid., June 14, 1947, p. 76.

[25] Ibid., p. 77.

[26] Jacoby, op. cit., p. 124.

[27] Quoted in Gabriel, op. cit., p. 61.

[28] U.S. Congress, p. 223.

[29] Mikesell and Chenery, op. cit., pp. 16-17.

[30] U.S. Congress, p. 4. Source: Department of the Interior; "b/d" stands for barrels per day.

[31] U.S. Congress, p. 12.

[32] Joseph E. Pogue, *Financial and Operating Data for 30 Oil Companies* (New York: The Chase Manhattan Bank, 1945), p. 18.

[33] Jacoby, op. cit., pp. 70-71.

[34] Raymond Vernon, *Storm Over the Multinationals: The Real Issues* (Cambridge, Massachusetts: Harvard University Press, 1977), p. 83; see also Raymond Vernon and Louis T. Wells, *Manager in the International Economy* (Englewood Cliffs, New Jersey: Prentice-Hall, 3rd ed., 1976), p. 244.

[35] *Oil and Gas Journal,* July 27, 1946, p. 131.

[36] Ibid., July 27, 1946, p. 131.

[37] J. Surnam, on November 9, 1945 quoting *Oil and Gas Journal,* February 15, 1947, p. 66.

[38] Ibid.

[39] *Oil and Gas Journal,* December 27, 1947, p. 144.

[40] Ibid., May 20, 1948, p. 125.

[41] Ibid., p. 125.

[42] U.S. Congress, p. 95.

[43] *Oil and Gas Journal,* May 27, 1948, p. 44.

[44] *Oil and Gas Journal,* July 1, 1958, p. 39. The actual figures were taken from the *Mineral Yearbooks* published by the U.S. Bureau of Mines.

[45] U.S. Congress, p. 114.

[46] *Oil and Gas Journal,* February 5, 1948, p. 46. "Company plans point to lower costs for crude imports from Middle East."

[47] *Oil and Gas Journal,* September 27, 1947, p. 61.

[48] Ibid., October 18, 1947, p. 84.

[49] 75% of the total number of reasons mentioned.

[50] 88% of the total number of reasons mentioned.

[51] See Chapter IV, Section IV, for a presentation of the vertical integration concept.

CHAPTER III

TEST OF A HYPOTHESIS ABOUT SOME INDEPENDENTS' FIRST ENTRIES ABROAD AFTER WORLD WAR II

Introduction

The purpose of this chapter is to submit to different statistical tests the hypothesis presented at the end of Chapter II. To that effect, this hypothesis has been reformulated in a testable way such that:

1) Given the threat by the majors represented by the import of cheap crude oil from the Middle East starting in 1946;
2) the lower the independent's self-sufficiency ratio (defined as its own production divided by its refinery runs), the quicker its initial entry into foreign exploration, the self-sufficiency ratio being a proxy for the firm's exposure to increased costs of finding oil in the United States compared with abroad;
3) the bigger the independent's size compared with the majors', the quicker its entry into foreign exploration, the relative size being a proxy for the firm's concern for oil reserves expansion and for the lumpiness of the first investment abroad; and
4) having an international experience prior to World War II will encourage an independent to move abroad faster than one without it.

In this chapter, the format followed is to examine each independent variable separately, and then all of them together, therefore going from the simplest to the more complex approach.

I. Average Year of Entry Abroad

Reading the literature dealing with the oil industry, one is under the impression that the independents went abroad in the middle 1950s, when they were invited either to join the Iranian Consortium or to bid for concessions in Libya and Venezuela. It is a wrong impression by several years. In fact, as shown in Exhibits 1 and 2, a majority of the independents were abroad in 1950.

Average years of entry abroad were computed, making the distinction between entry abroad including Canada and entry outside the U.S. and Canada. The rationale is that, in the late 1940s and the early 1950s, many oil companies

Exhibit 1
CHRONOLOGY OF THE INDEPENDENTS' ENTRY ABROAD

22 Companies	first entry abroad after WWII including Canada	outside U.S. and Canada
AMERADA	1948	1952*
ASHLAND	1948	1948
ATLANTIC	1945	1945
CITIES SERVICE	1946	1946
CONTINENTAL	1947	1952*
GETTY	1948	1949*
MARATHON	1949	1952*
MURPHY	1951	1956*
OCCIDENTAL	1961	1961
PHILLIPS	1944	1944
PURE OIL	1956	1956
RICHFIELD	1946	1946
SIGNAL	1948	1948
SINCLAIR	1945	1945
SKELLY	1948	1958*
STANDARD INDIANA	1948	1948
STANDARD OHIO	1945	1945
SUN OIL	1944	1956*
SUNRAY	1948	1957*
SUPERIOR	1947	1947
TIDEWATER	1949	1956*
UNION OIL OF CALIFORNIA	1947	1947
Average year of entry	1948	1950

* if the dates in the two columns are different.

SOURCE: Companies' *Annual Reports.*

did not consider Canada as a foreign country. The independents were exploring outside the U.S. in average by 1947 and outside the U.S. and Canada by 1950.[1]

II. The Evidence of Threat by Competitors

The proxy for threat is the imports of crude oil from the Middle East divided by the U.S. production. This normalization was designed to remove part of the time factor from the imports time series, thus leaving just the acceleration of the imports trend over that of domestic production.

The ratios were computed from data published by the U.S. Bureau of Mines in the *Mineral Yearbook* for the years 1945 to 1959. Two numerators—imports from the Middle East and worldwide imports—and two denominators—U.S. production and U.S. consumption—were considered. As shown in Exhibit 4, whatever the definition used, all proxies of threat are still strongly correlated with time, even after normalization. All the linear correlations produced coefficients higher than .9—as shown in Exhibit 3—and a logarithmic fit worked

Exhibit 2
Distribution of the Independents' Entries Abroad

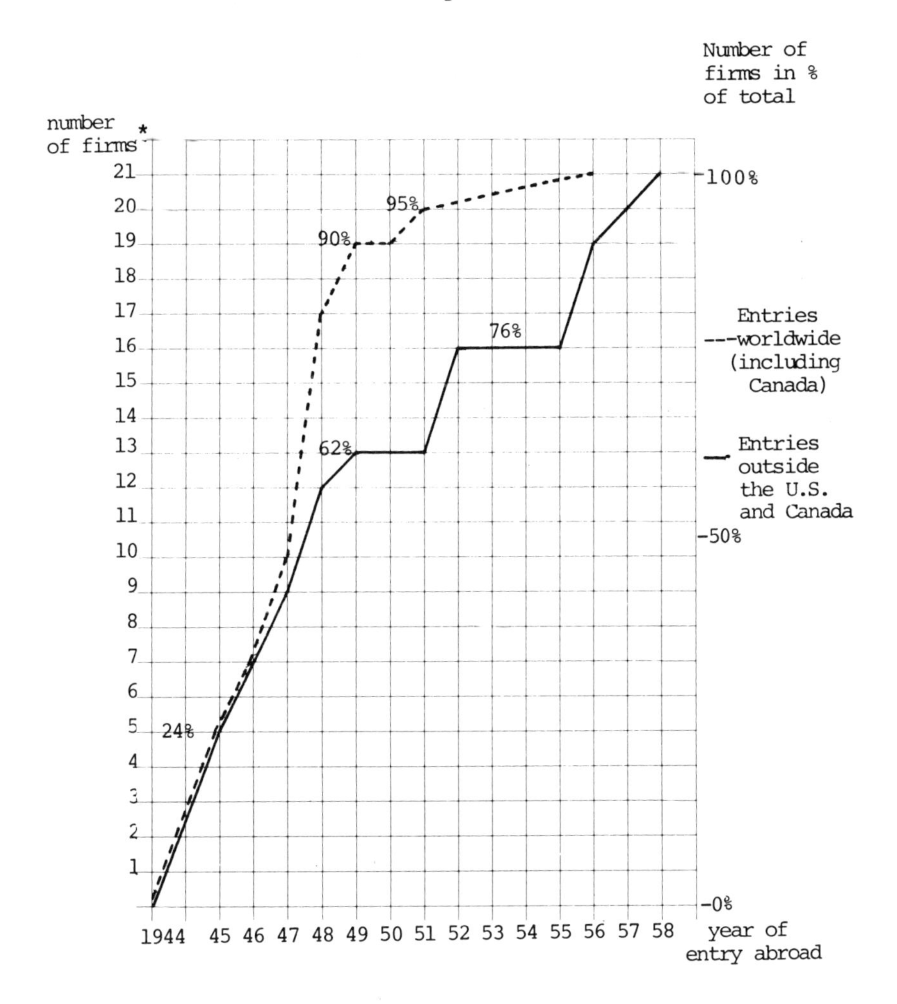

*Excluding Occidental

SOURCE: Exhibit 1.

Exhibit 3
CORRELATION BETWEEN PROXIES OF THREAT AND TIME
(1946-59)

	IMPORTS	
	Middle East	Worldwide
U.S. production	.95	.97
U.S. consumption	.91	.95

better than a linear fit in two cases.

Because the threat was on the increase throughout the period, its measures are also strongly correlated with the time-series formed with the independents' dates of entry outside North America. Correlation coefficients are .97, taking imports from the Middle East (normalized by U.S. production), and .99, taking worldwide imports (normalized the same way).

When the time element is removed by taking the variation of the ratio (variation measured in percentage points of the ratio) from one year to another instead of the ratio itself, correlations between acceleration of threat and the companies' dates of entry abroad are lower. Yet they suggest links between entries abroad and threat by imports. The correlation coefficient is .10 (significant only at the 35% level using the t test) taking imports from the Middle East. But it is .32 (significant below the 10% level) taking worldwide imports, thus suggesting that the independents may have been more sensitive to the acceleration of the total imports than just of the imports from the Middle East.

III. The Influence of Size

The size of each independent was measured by its total dollar income normalized by the sales of the largest major–Exxon.

The relative size data were derived from total income figures published in companies' annual reports. Because it was not possible to get income from oil activity alone, total income was taken as the closest proxy. Since some of the independents began diversifying into petrochemicals and other unrelated businesses, this approximation introduced a flaw in the data base. However, most of the surveyed companies went abroad while their main activity was in petroleum.

To probe whether the biggest among the independents were the first to go abroad, I used the t test for two means. The proper way would have been to compare, for each year, the average size of those that went abroad with the size of those that did not go abroad that year. Because of the small sample size,[2]

Exhibit 4
Relationship between Threat by Imports and Year of Entry Abroad

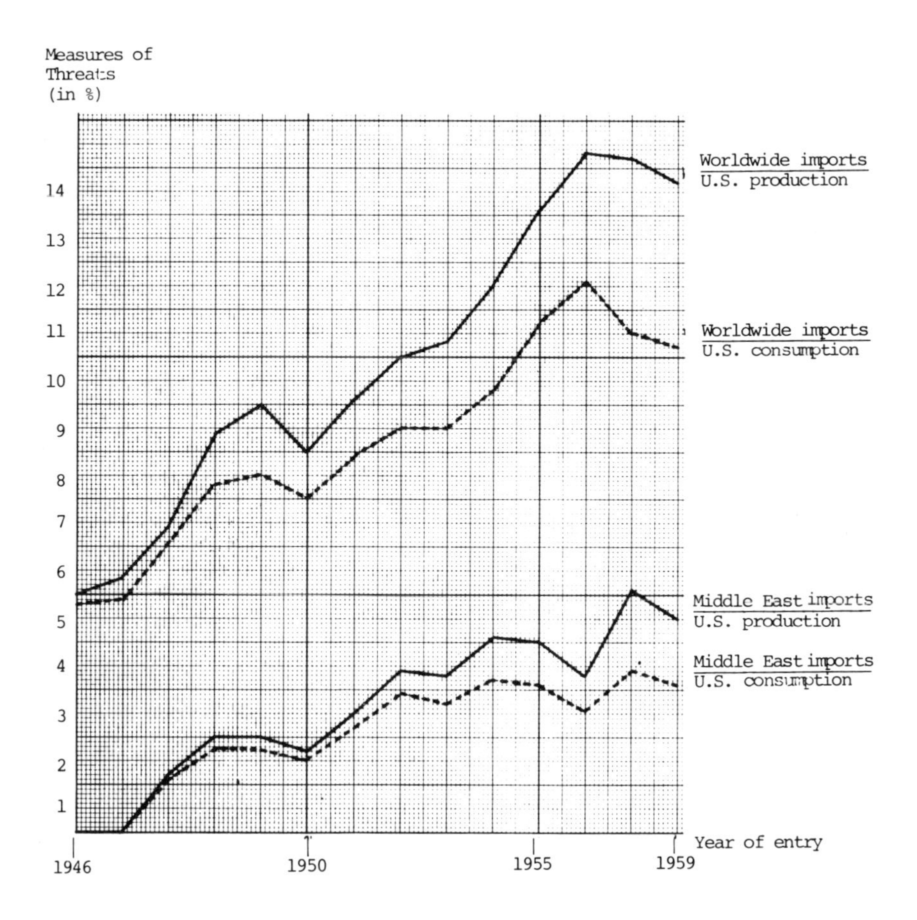

SOURCE: U.S. Bureau of Mines' *Mineral Yearbooks.*

however, such comparisons would not have been meaningful.

In order to increase the degree of freedom, I did not take for each year just the size of the companies going abroad that particular year, but also the size in that same year of the independents that went abroad in previous years.[3] Because of this convention, it is like checking whether the "international" independents' relative size was higher each year than the still "domestic" independents'.

Such a procedure assumes, however, a stability of the relative size throughout the period. Accordingly, a Spearman correlation between the size ranks in 1945 and the size ranks in 1959 produced a coefficient of .86, significant far below the 0.5% statistical level.

The results, summarized in Exhibit 5, tend to support the hypothesis, although not at very high statistical levels. The size ratio of the "domestic" independents to the "international" ones was, however, always below 1 and averaged .68 plus or minus 10%.

Exhibit 5
SIZE OF FIRM AND PROPENSITY TO GO ABROAD
U.S. Independent Oil Companies
1945 through 1955

	Independents					
	Venturing Abroad		Not Venturing Abroad			
Year	Size [a] (in % of Exxon's)	# of Firms	Size [b] (in % of Exxon's)	# of Firms	Probability That Difference in Size Occurred By Chance	Size Ratio [b/a]
1945	15.17	3	9.99	17	20%	.66
1946	14.10	6	8.57	14	10%	.61
1947	10.85	8	8.53	12	30%	.79
1948	11.28	11	6.50	9	10%	.58
1949	11.50	12	7.85	8	20%	.68
1950	11.93	12	7.98	8	20%	.67
1951	11.72	12	7.41	8	15%	.63
1952	10.46	15	8.30	5	35%	.79
1953	10.87	15	7.28	6	25%	.67
1954	8.05	15	5.33	6	22%	.66
1955	7.81	15	5.65	5	25%	.72

SOURCE: U.S. Independents' and Exxon's *Annual Reports.*

The influence of size on the decision to go abroad was confirmed by a Spearman correlation between the size ranks in 1945 and the entry ranks. The test showed that the first independent to go abroad tended to be the biggest among them, although some exceptions were noted.[4] A Pearson correlation helped to qualify this point. As shown in Exhibit 6, several small independents

Exhibit 6
Relationship* Between Size and Year of Entry Into Exploration Outside the U.S. and Canada**
U.S. Independent Oil Companies, 1945 through 1958

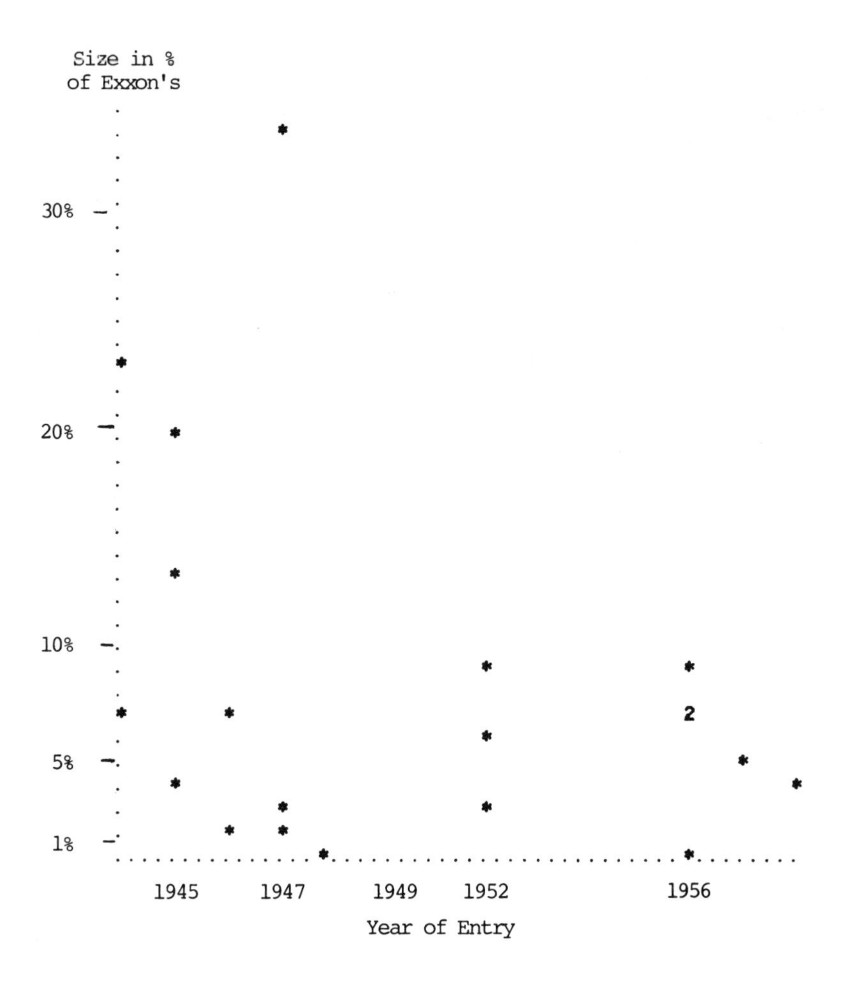

* Correlation coefficient: -.33%, significant below 10%.

** Diagram produced by the program "PLOTYX" of Harvard Business School's computer system.

SOURCE: U.S. independents' *Annual Reports*.

[A closely similar pattern is observed when entries into Canada are included.]

managed to go abroad quickly. All the latecomers were small, however, while all the big firms went abroad quickly. The negative correlation coefficient confirmed that the bigger the size, the quicker the entry abroad.

IV. The Influence of the Vertical Integration's Balance

The same test as for size was repeated on companies' self-sufficiency ratio[5] to see whether, each year, "international" independents had self-sufficiency ratios lower in average than still-"domestic" independents. The results are shown in Exhibit 7. They tend to support the hypothesis that the degree of vertical integration between production and refining was a key factor in the independents' decision to go exploring abroad. This is particularly evident for the companies going outside the U.S. and Canada between 1948 and 1951.

Exhibit 7
FIRM'S CRUDE POSITION AND PROPENSITY TO GO ABROAD (OUTSIDE THE U.S. AND CANADA) U.S. INDEPENDENT OIL COMPANIES, 1945 THROUGH 1955

	Independents				
	Venturing Abroad		Not Venturing Abroad		Probability That Difference in Degree of Self-Sufficiency Occurred By Chance
Year	Self-Sufficiency (production/ refinery runs)	# of Firms	Self-sufficiency	# of Firms	
1945	.62	3	1.36	11	20%
1946	.53	6	1.31	10	10%
1947	.55	7	1.33	9	10%
1948	.49	9	1.42	7	2.5%
1949	.45	9	1.17	7	1%
1950	.43	9	1.20	7	0.5%
1951	.46	9	1.29	7	0.5%
1952	.69	11	1.06	5	15%
1953	.67	11	.98	5	20%
1954	.67	11	.96	5	25%
1955	.64	11	.74	5	40%

SOURCE: U.S. independents' *Annual Reports.*

Not only were the average self-sufficiency ratios significantly different, but until 1951 the international independents' average ratio was decreasing while the domestic independents' one was more-or-less constant. In other words, the independents going abroad were the "crude short" ones, which foresaw a worsening of their shortage, while those staying in the U.S. were the "crude long" ones, which did not have to worry about crude oil supply.

This conclusion is confirmed by the correlation between the dates of entry and the companies' self-sufficiency ratios that year, which produced a positive and statistically significant coefficient (see Exhibit 8). I conclude, therefore, that the shorter the crude position, the greater the incentive to go abroad.

As shown in Exhibit 9, the correlation between size and crude position confirmed previous findings. Independents going abroad in the 1940s tended to be big and crude short, whereas those following in the 1950s tended to be small and crude sufficient. The linear correlation between the companies' sizes and crude positions had a negative sign, but is not very significant by any statistical standard. That suggests a weak autocorrelation between the two variables.

V. Influence of International Experience Prior to World War II

From the data presented in Appendix 4, the independents were divided into two groups: those firms involved abroad before World War II, and those not. The hypothesis formulated in the introduction of this chapter about the influence of a previous experience after 1945 was not supported. A t test showed no difference in the dates of entry into foreign exploration between the firms having been involved abroad and those not, taking entries either outside the U.S. or outside the U.S. and Canada.

In consequence, no independent variable representing the independent's previous experience abroad was included in the regression test. Doing so through a dummy variable coded 0 for no previous experience abroad and 1 for previous experience would have reduced further a degree of freedom already low. In addition, as it will be seen in the next section, just one variable—the proxy for threat—explains the major part of the coefficient of determination, thus leaving a limited explanation power to the two other variables linked with the companies' behavior. Adding another weak one would not have improved the regression analysis.

VI. Regression Test

The hypothesis formulated in the introduction was finally submitted to a multiple linear regression test incorporating all the independent variables examined separately. The regression equation tested was (with the expected signs):

$$Y = a + bX_1 - cX_2 + dX_3$$

where:

1) the dependent variable Y is the number of the year after the first import of crude oil from the Middle East (1946) that a given

Exhibit 8
Relationship between Crude Position and Year of Entry Into Exploration Outside the U.S. and Canada*
U.S. Independent Oil Companies, 1945 through 1958

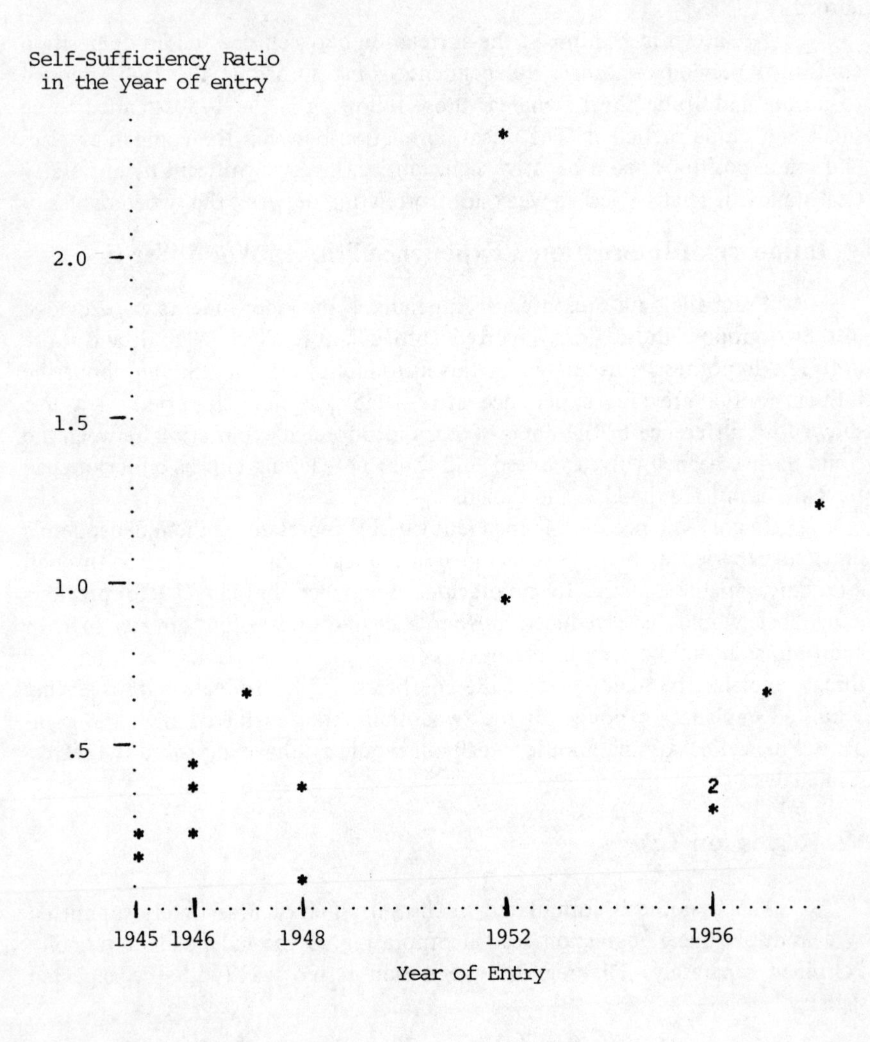

*Correlation coefficient: -.34, significant at 10%.
SOURCE: U.S. independents' *Annual Reports.*
[A similar pattern is observed when entries into Canada are included.]

Exhibit 9
Relationship between Relative Size and Self-Sufficiency in the Year of Entry Outside the U.S. and Canada U.S. Independent Oil Companies, 1945 through 1958*

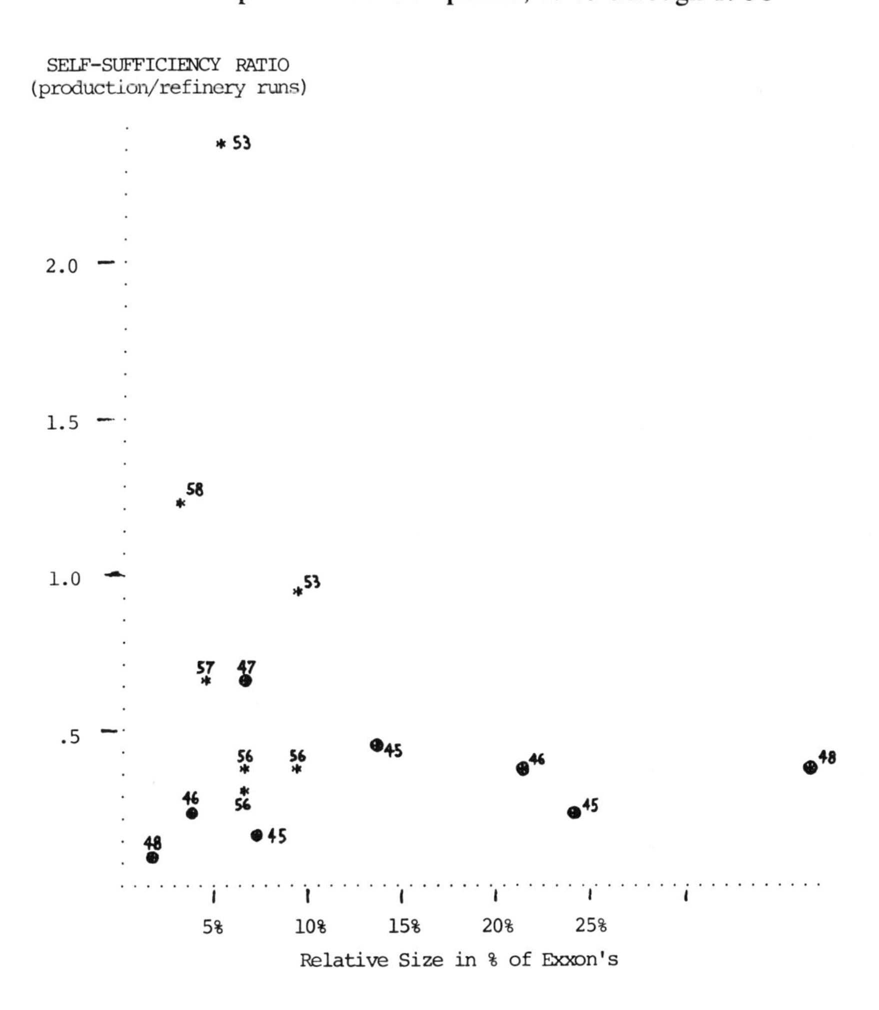

* Numbers are the last two digits of the company's year of entry.

1. Correlation coefficient: -.23 (significant at only 20%)

[A similar pattern is observed when entries into Canada are included.]

independent oil company went exploring abroad (including Canada);

2) X_1 = independent's self-sufficiency ratio (that is, its own production divided by its total crude refinery runs) in the year of entry abroad;
3) X_2 = size of the independent (measured by its total dollar income normalized by the size of the largest major) in the year of entry abroad; and
4) X_3 = imports of crude oil from the Middle East (in percentage of the U.S. production) in the year a given independent went abroad.

Because the hypothesis deals with a concept of threat represented by imports of crude oil from the Middle East, and because those started in 1946, the entries abroad having taken place during World War II for the purpose of oil exploration were excluded from the regression. This was the case of four independents: Phillips, Sinclair, Standard of Ohio, and Sun Oil. However, those firms going abroad just before the war, if any, were kept in the sample. The nonintegrated producers without any self-sufficiency problem were excluded. Such was the case for Amerada, Getty, Murphy, Signal, and Superior. Pure Oil was excluded because, with an entry abroad as late as 1956, though associated with a low self-sufficiency ratio, it caused a wrong sign in the regression equation. Accordingly, excluding this independent from the sample, though lowering slightly the R^2, contributed in giving the correct signs for all the independent variables. Occidental was excluded because it went abroad after the period under consideration for the regression.

The test's results are summarized in Exhibit 10.

Exhibit 10
RESULTS OF THE REGRESSION RUN

	Regression Coefficient	Probability for sign of β to be correct	R^2
Self-sufficiency (X_1)	.0013*	.75	.232
Relative size (X_2)	-.0023	.57	.002
$\frac{\text{Imports from Middle East}}{\text{U.S. production}}$ (X_3)	1.27	1.	.684
Constant	1.33	1.	
Year of entry abroad (Y)			.918

F test = 10.09 (significant below 0.1%)
Durbin-Watson test = 1.77 (significant at 2.5%)
Residual standard deviation: .308
Number of observations: 11

The hypothesis is statistically supported. The coefficient of regression is high and significant; the signs of the independent variables are all correct; the residual standard deviation is low; and the Durbin-Watson test shows no serial correlation among the independent variables.

The regression test presents, however, two drawbacks that must be noted. First, 75% of the multiple R^2 is due to only one variable–imports from the Middle East normalized by U.S. production. This result is not unexpected, for the high correlation of the measure of threat with time has been mentioned earlier; but it probably gives to this variable an excessive explanatory power by comparison with the two others, which are relatively time-free. Second, only half of the independents under study are covered by the regression analysis.

In Appendix 3, several variants of the base regression are tested, excluding entry into Canada, measuring the independent variables in the year preceding the entry, and measuring threat by the level of total imports of just those from the Middle East.

Conclusion

In this chapter, I tested an explanation of the independents' entries abroad immediately after World War II. The main finding is that several factors were at work. Rising imports of cheap crude oils from the Middle East triggered a defensive mechanism on the part of the independents. They were thus compelled to go abroad in order to match the edge of their bigger competitors. At the same time, greater costs of exploration in the U.S. compared with abroad, associated with lower productivity of wells in case of discovery, put foreign exploration at a profit advantage over a purely domestic strategy. In that respect, the crude short companies were more likely to consider the foreign alternative than the crude sufficient ones, for the former had a more pressing need to feed their refineries. In addition, by going abroad, an independent was able to reduce the risk associated with exploration, since the likelihood of a better productivity was greater abroad than in the U.S. The tests also showed that lumpiness of the first foreign investment was a barrier to entry; thus, a company's size was a factor. Yet, several small firms managed through joint ventures with the big independents[6] to share the financial burden of the initial investment abroad. There was, however, no evidence that previous international experience caused some independents to move more quickly abroad than the newcomers to international business.

A test dealing with the subsequent decisions in foreign exploration over the period 1945-1976 across countries showed that the independents tended to follow the others in the implementation of their exploration strategy abroad.[7] In other words, whereas the first decision was triggered by a profit-maximization and risk-minimization behavior, with the imports of Middle

Eastern oil being the triggering mechanism, the subsequent decisions of geographical allocation of exploration resources seem to have been oligopolistic reactions to competitive threats by the fellow independents.

NOTES

[1] A t test for two means showed that average years of entry outside the U.S. and outside the U.S. and Canada were significantly different (at the 5% level). Actually such a test is meaningless, for I am dealing with a population and not with samples. It may give some readers more confidence in the finding, however, although it will suffice to say, in this case, that the two dates of entry are "different." The same reservation applies for the subsequent uses of the t test.

[2] An average of 1.5 entries per year between 1944 and 1958.

[3] Because, in that case, the t test is applied no more on a population but on a data generation process such that data (sizes here) are measured the same way each year, its use is appropriate.

[4] The correlation coefficient was .27 (significant slightly above the 10% level) when only entries outside the U.S. and Canada were taken into account. But it was .41 (significant below 5%) when entries into Canada were also considered.

[5] The self-sufficiency ratios, defined as own production divided by refinery runs, were computed from production and refinery figures published in annual reports. When the independent was only a producer—like Amerada, Getty, Murphy, Signal and Superior—no ratio was calculated, although it should have been, in principle, infinite. In most of the cases production data are those of crude oil only. When it was not possible to disaggregate these figures from natural gas liquids production, the self-sufficiency ratios were computed as such.

[6] See Chapter VI, Section I for further evidence.

[7] See Appendix 5.

Part II

The Subsequent Decisions to Invest and Divest

CHAPTER IV

A FRAMEWORK FOR THE ANALYSIS OF INTERNATIONAL STRATEGIES OVER TIME

Introduction

From here on, there is a change of focus in the research. The rest of this study will be concerned with all the subsequent decisions the independents made in the course of their growth abroad.

Chapter IV presents the model that I used to sort out and analyze the twenty-two independents' international strategies. But first, contributions in general management, international business, and industrial organization are acknowledged in three different sections.

I. Business Policy Literature on Strategy

The concept of corporate strategy occupies a central position in the study of business policy or general management. The term "strategy," however, has been used and described in a multitude of ways. To simplify, there are two basically antithetic concepts of strategy. The first defines strategy as a "conscious plan"; the second defines it as an "organizational outcome."

Chandler has pioneered the first definition: "Strategy can be defined as the determination of the basic long-term goals and objectives of an enterprise, and the adoption of courses of action and the allocation of resources necessary for carrying out these goals."[1] With Andrews, the identification of the corporation's role in this "conscious plan" is more strongly emphasized: "The strategic decision is the one that determines the nature of the business in which a company is to engage and the kind of company it is to be."[2] For Scott, the term strategy includes: "a concept of how to compete in an industry . . . a statement of specific goals against which progress can be measured (e.g., a target market share to be attained, a target rate of growth in sales, a target rate of ROI) . . . and a timed sequence of conditional moves for deploying skills and resources with a view to attaining one's objectives."[3] Beyond the nuances, in all these three definitions, strategy is made of two equally essential aspects: formulation and implementation.

A competitive and radically different concept of strategy is derived from the behavioral theory of the firm developed by Cyert and March at Carnegie-Mellon University[4] (and applied by Aharoni and Bower at Harvard Business School).[5]

Cyert and March reject the institutional view of an organization with a consciously selected, consistent set of goals. Rather, the goals of an organization are the "outcome" of a series of bargains among coalitions of individuals. "The goals of a business firm are a series of more or less independent constraints imposed on the organization through a process of bargaining among potential coalition members and elaborated over time in response to short-run pressures. Goals arise in such a form because the firm is, in fact, a coalition of participants with disparate demands, changing foci of attention, and limited ability to attend to all organizational problems simultaneously."[6]

By considering alternative definitions of strategy, I am not trying to decide which concept of strategy is "correct." The purpose is rather to make clear under which alternative definition of strategy this study will be conducted. Since, at this stage, I am more concerned with strategic outcomes than strategic processes, this analysis will be placed within the perspective of Harvard's view of strategy and more specifically will adhere to Chandler's definition.

So far the research in business policy has not focused on the patterns of strategy over a long period of time. That does not mean that the time dimension is completely missing. But at the best, comparative statics or "spread" statics have been used to analyze structural changes. Illustrative of the former approach is Rumelt's study of diversification through three bench marks set in 1949, 1959, and 1969, with all the sensitivity of the findings to such arbitrary dates that choice implies.[7] Illustrative of the second approach is Chandler's causal linking of strategy and structure, where he spotted changes in structure and then, looking backward, found that changes in strategy had closely preceded.[8] Chandler, in turn, inspired Scott's model of stages of corporate development.

II. International Business Literature on International Business Strategy

This section has for its purpose to review studies dealing expressly with international strategy as a whole. By that I mean those analyses claiming that they considered the overall strategies abroad of multinational firms and not merely isolated foreign direct investment decision processes.

Following this definition, one is obliged to conclude that the theory of international business strategy is just emerging from infancy in contrast to substantial portions of the international business field.

Brooke and Remmers' contribution,[9] for example, boils down to a distinction between "defensive" and "aggressive" strategy, whose usefulness is limited for my purpose. The drawback comes less from the sample of firms they are dealing with–European manufacturers–than from their narrow definition of international strategy, which is focusing solely on the first foreign direct investment decision. It is perfectly valid to consider such a decision as strategic. But doing so results in leaving out all the flow of subsequent strategic decisions that follow the first entry abroad.

Stopford and Wells,[10] as well as Brooke and Remmers, tried to overcome this problem, but they raised another one. In both cases, their proxy for international strategy covers only one aspect of it: the organization designed to implement the international strategy.

I do not mean the approach is wrong; the authors are perfectly clear about it: "The study deals with the ways in which multinational enterprises have altered their structure as they have developed new and more complex strategies."[11] Although the descriptive conceptual framework they developed contributed greatly to the advance of the field, there is a need now to go beyond this kind of analysis.

Indeed, the relationship between strategy orientation and organizational structure is now well recognized as a result of Chandler's pioneering work.[12] In fact, it is so well accepted that Stopford and Wells go as far as to characterize international companies by the type of structure they adopt. If this is correct in the first approximation and for certain purposes, strategy has, nevertheless, a broader meaning (as recalled in the previous section), which must be explored.

III. Industrial Organization Literature on Strategic Groups

As I have done for the other bodies of knowledge, I will survey in this section only those works of industrial organization literature that have some bearing on the subsequent analysis.

Industrial organization is concerned with the explanation of industry performance. It differs from traditional micro-economic theory in the set of independent variables it considers and in its concern for concrete real-world cases.

Since Hymer's thesis,[13] the influence of industrial organization on international business has been growing. Decreasing barriers to entry have already been mentioned as one factor for the move of the independents abroad; and the key concept of follow-the-leader has been pointed out. A still less widely accepted approach could also be borrowed from industrial organization: the notion of strategic groups.

The whole argument of industrial organization economists rests on the traditional economic view of the firm—all firms in a given industry are the same except for size. It is the assumption of symmetry. In fact, the notion of a unique industry is in many cases inappropriate. Common observation at the firm level suggests that the firms in an industry often differ from one another in their degree of vertical integration or diversification, the extent of their involvement abroad, and so forth. In other words, strategies differ among firms in an industry. To deal with asymmetric industries, Hunt, and then Porter, developed the concept of the strategic group.[14] An industry thus may consist of groups of firms, each group composed of firms that are highly symmetric, that is, having similar strategies. This concept is crucial for this research, as it will be suggested in the next section.

IV. The Conceptual Framework That Will Be Used

The analytical framework I propose to use for analyzing the independents' international strategies I have called "multinationalization through verticalization." It draws from the concept of vertical integration, but also from the concept of stages of development of strategy.

Stages of development applied to structure is not a new concept. Scott developed it in the early 1960s, and it is at the core of Stopford and Wells' work: "The organizational structure of a firm can be thought of as evolving in a series of stages. Each stage is a modification or adaptation of the structure in the previous stage. The expertise and experience of the enterprise with one structure provide the building blocks for future structures."[15] Strategy per se, however, has been left out of the model, although one can argue that "structure is strategy." Indeed, Stopford and Wells added that: "The strategy of the firm undergoes a similar process";[16] but they did not elaborate on it, for their interest was on structure.

Mira Wilkins, however, in her historical description of the U.S. multinationals' growth abroad, applied this concept of stages of development to international strategy.[17] Although based on observations, her model dealing with the petroleum companies stays at too broad a level of generalization to be taken as an exploratory model for this research. She makes the distinction merely between three stages: export of petroleum products; production abroad; production in one country for selling in another. Furthermore, it has a methodological shortcoming that impinges on its use. The model is supported by cross-sectional evidence (that is, different companies at different times), but not by longitudinal evidence (same companies at different times).

Vertical integration is an outstanding characteristic among the large firms in the domestic oil industry, which has been studied extensively. The consensus is that vertical integration in the U.S. petroleum industry represents a competitive response to the drawbacks of market contracting. If intermediate products' contractual purchases and sales could be made without friction, there would be no need for internalizing transactions within a vertical structure. By being integrated, an oil company is able to schedule production, manufacturing, and inventories more efficiently, to coordinate complementary investments in the different stages of the oil business more exactly, and therefore to adapt to changing economic conditions more effectively.[18]

Although vertical integration presents economic advantages of its own, it has been often implemented as a defensive response to oligopolistic market conditions. Relating the integration propensity of the oil firms to the nature of the competition, DeChazeau and Kahn concluded that: "If no companies were vertically integrated, it would be less urgent for any of them to be."[19] In this research, however, I viewed vertical integration as an autonomous process rather than an imitation one dictated by the competitors' behavior.

If scholars recognized that the integration of the major companies extended across national frontiers,[20] it did not strike as a feature of some of the independents. Accordingly, in 1968, Vernon noted that "some companies, especially the newer and smaller ones in the international field, acquired crude before they had developed outlets for it."[21] I assumed, however, that, for the same reason they integrated domestically and the majors integrated abroad, the independents will follow the same pattern in the long run.

Thus, the analytical model I used was based on two premises: (1) that there are different phases in the petroleum business–mainly exploration, production, refining, and marketing–in which the oil companies can be involved, but do not necessarily have to be; and (2) that, starting with foreign exploration as the first step, a process of vertical integration abroad will occur in the chronological order depicted in Exhibit 1.

Actually, this sequence is not linear, as some choices are open to Stage 3 and Stage 5. The concept proposed is therefore better represented by a diagram, as in Exhibit 2.

Transportation is not dealt with in this research, for an oil company does not need to be active abroad in order to be integrated into transportation. Being an importer of foreign crude oil would be a sufficient incentive. This stage was therefore left out of the model.

From this representation of different stages of multinationalization through verticalization it is easy to derive the concept of strategic groups, as described in Exhibit 3.

Although the stage-of-development concept integrates the time dimension, the notion of strategic groups given by Exhibit 3 is rather static. A company will be sorted out in a given strategic group, say Group 4, if somewhere during the surveyed period it went up to Stage 4: refining abroad. But this classification says nothing about the strategy's dynamic over time, that is, about the changes of strategy and the time lags between them. This dynamic is illustrated in Exhibit 4 for the cases of two companies that implemented a complete vertical integration abroad. Although the resulting strategic positions are identical, the strategies' time patterns are sensibly different.

I used this analytical model both for gathering published data on the companies and as a guide for my interviews with executives.

Exhibit 1 – STAGES OF DEVELOPMENT OF INTERNATIONAL STRATEGY

	Stage 1	Stage 2	Stage 3	Stage 4	Stage 5	Stage 6	Stage 7
Strategic Decisions	Exploring abroad	Developing abroad if discovery Producing abroad if commercial discovery	Selling foreign crude on international market (3) or importing it back in the U.S. (3A)	Refining its foreign crude abroad through either own refinery (ies) or refining arrangements	Selling its foreign refined products at refinery's doors (5) or importing them back in the U.S. (5A)	Distributing abroad its own refined products at wholesale level	Same as 6, but selling at retail level

Exhibit 2—Stages of Development of International Strategy

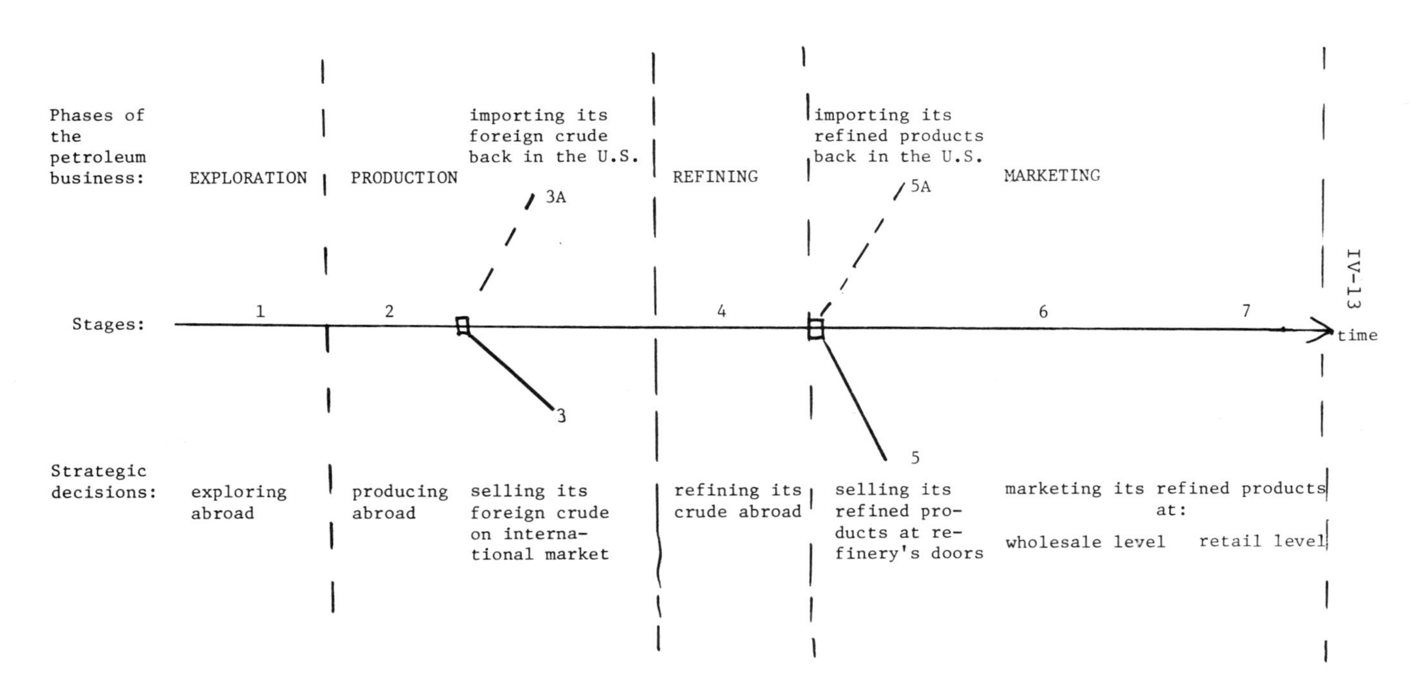

Exhibit 3—Strategic Groups

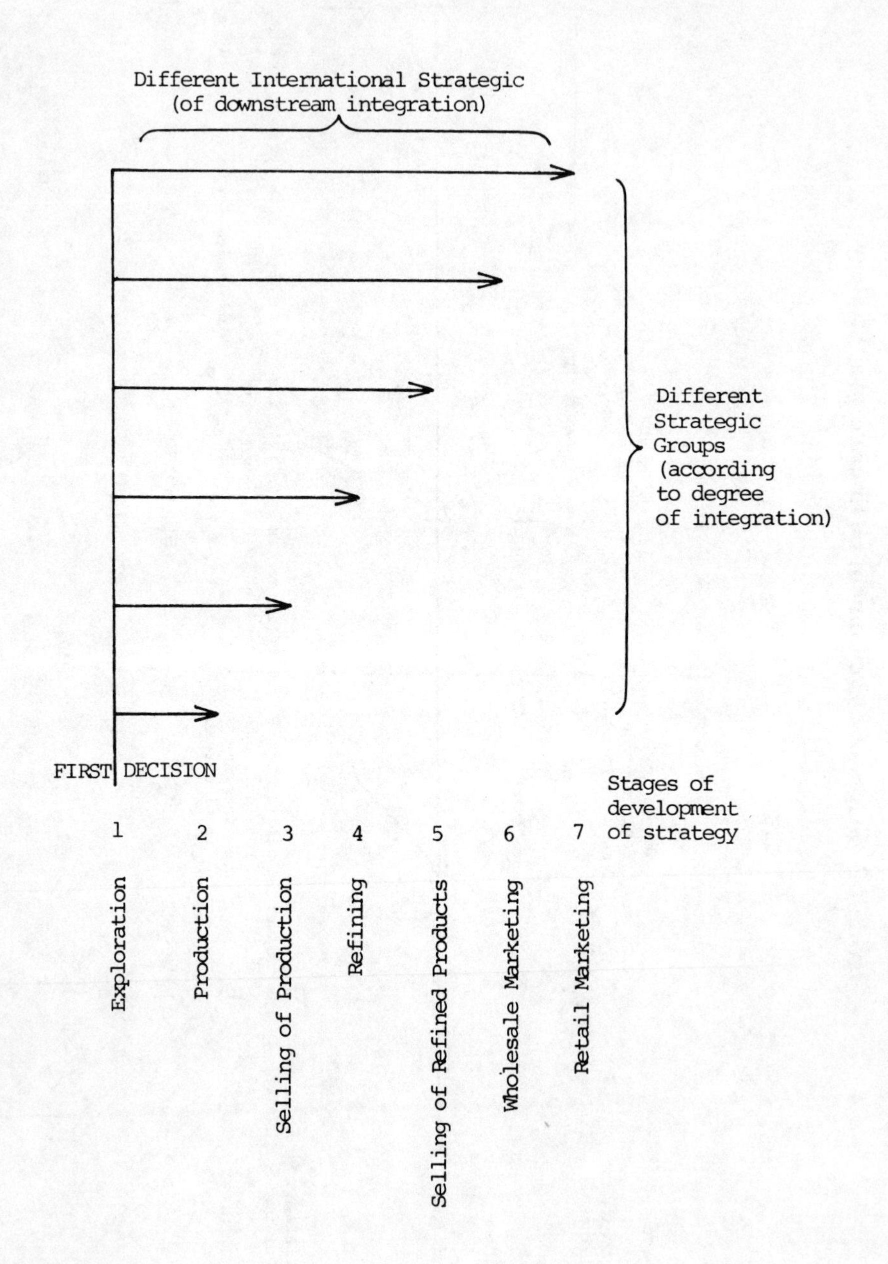

Exhibit 4—Evolution of International Strategies Over Time

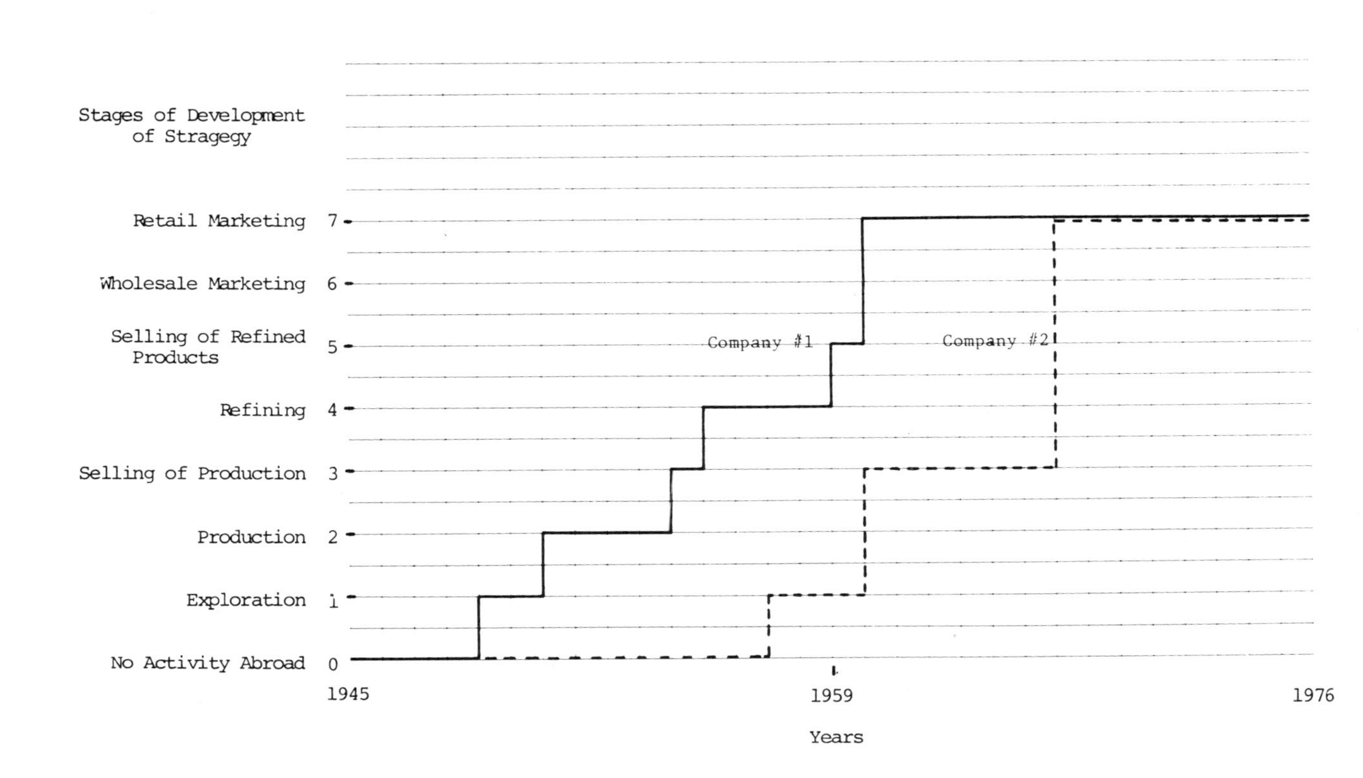

NOTES

[1] Alfred D. Chandler, Jr., *Strategy and Structure* (Cambridge, Massachusetts: The MIT Press, 1962), p.

[2] R. C. Christensen, K. R. Andrews, and J. L. Bower, *Business Policy–Text and Cases* (Homewood, Illinois: Richard D. Irwin, 1973), p. 9.

[3] John H. McArthur and Bruce R. Scott, *Industrial Planning in France* (Boston: Division of Research, Graduate School of Business Administration, Harvard University, 1969), pp. 113, 115.

[4] Richard M. Cyert and James March, *A Behavioral Theory of the Firm* (Englewood Cliffs, New Jersey: Prentice-Hall, 1963).

[5] Yair Aharoni, *The Foreign Investment Decision Process* (Boston: Division of Research, Graduate School of Business Administration, Harvard University, 1966); and Joseph L. Bower, *Managing the Resource Allocation Process* (Boston: Division of Research, Graduate School of Business Administration, Harvard University, 1970).

[6] Cyert and March, *A Behavioral Theory,* p. 30.

[7] Richard P. Rumelt, *Strategy, Structure and Economic Performance* (Boston: Division of Research, Graduate School of Business Administration, Harvard University, 1974).

[8] Chandler, *Strategy and Structure.*

[9] Richard Z. Brooke and H. Lee Remmers, *The Strategy of Multinational Enterprise* (London: Longmans, 1971).

[10] John M. Stopford and Louis T. Wells, *Managing the Multinational Enterprise* (New York: Basic Books, 1972).

[11] Stopford and Wells, op. cit., p. 4.

[12] Chandler, *Strategy and Structure.*

[13] Hymer, *International Operations.*

[14] Michael Hunt, "Competition in the Major Home Appliance Industry, 1960-1970" (Unpublished doctoral dissertation, Harvard University, 1972); H. H. Newman, "Strategic Groups and the Structure-Performance Relationship" (Unpublished doctoral dissertation, Harvard University, 1973); Michael E. Porter, *Interbrand Choices, Strategy and Bilateral Market Power* (Cambridge, Massachusetts: Harvard University Press, 1976).

[15] Stopford and Wells, op. cit., p. 10.

[16] Ibid.

[17] Mira Wilkins, *The Maturing of Multinational Enterprise–Business Abroad from 1914 to 1970* (Cambridge, Massachusetts: Harvard University Press, 1974).

[18] See David J. Teece, *Vertical Integration and Vertical Divestiture in the U.S. Petroleum Industry* (Stanford, California: Graduate School of Business Administration, Stanford University, Research Paper no. 300, 1976); see also McLean and Haig, *Growth of Inte-*

grated Oil Companies.

[19] Melvin C. DeChazeau and Alfred E. Kahn, *Integration and Competition in the Petroleum Industry* (New Haven, Connecticut: Yale University Press, 1959).

[20] See Edith T. Penrose, *The Large International Firm in Developing Countries: The International Petroleum Industry* (London: Allen and Unwin, 1968), p. 46ff.; and Vernon, *Manager,* p. 28ff.

[21] Raymond Vernon, *Manager in the International Economy* (Englewood Cliffs, New Jersey: Prentice-Hall, 1968, 1st ed.), p. 135.

CHAPTER V

THE MODEL OF "MULTINATIONALIZATION THROUGH VERTICAL INTEGRATION" TESTED

Introduction

The model developed in Chapter IV is tested on two periods: before World War II and after. Although this study deals essentially with the post-War period, the model was applied to the pre-War period on the assumption that what the independents did then could have some bearing on their behavior after the war.

If one accepts the view of strategy as a chain of outcomes along a time vector, it matters considerably when the starting point is. As is shown in Exhibit 1, whether the first entry is A (before WWII) or B (after) results in a longer or shorter experience in strategy abroad. Not only has the company had more or less experience in dealing with foreign issues, but it may be assumed that the pre-War experience would have influenced the post-War strategy in terms of either alternatives considered or implementation.

Exhibit 1
LENGTH OF STRATEGIC EXPERIENCE

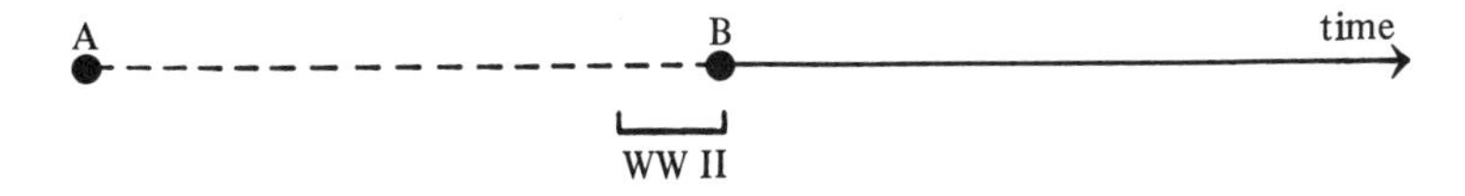

I. The Model Applied to the Pre-World War II Period

Twelve out of the twenty-two independents studied went abroad before World War II, thus each accumulating a previous experience in international business. Since the detailed descriptions of these experiences are in the Appendix,[1] this section will be concerned only with the main findings, which are summarized in Exhibits 2 and 3.

The model of stages of development of international strategy, which assumes a process of vertical integration abroad starting with exploration and then following the natural flow of crude oil up to marketing, was partly supported.

Eight out of the twelve independents involved abroad started with exploration, although two of them simply stopped at that stage. The remaining

Exhibit 2
INDEPENDENTS' INTERNATIONAL STRATEGIES BEFORE WORLD WAR II

Company	Evolution of International Strategy	Emphasis of Strategy
Amerada	1921: exploration (Alberta, Canada) 1926: Ibid. (Venezuela)	Strategy of having a portfolio of undeveloped acreage.
Ashland	no involvement abroad	
Atlantic	1919: exploration and production (Mexico) 1922: export business and direct marketing 1925: refining (Mexico) 1926: exploration (Venezuela)	Exploration, production & refining in Western Hemisphere Marketing world-wide
Cities Service	1918: exploration (Mexico) 1922: export business (Europe) 1920s: production (Mexico) 1930s: direct marketing (Argentina)	Oil business side line to domestic utility business
Continental	1920: exploration (Mexico) late 1920s: production (Mexico) 1930s: export business	Exploration close to the U.S.; no direct marketing Just export
Getty	no involvement abroad	
Marathon (Ohio Oil)	1926: exploration (Mexico) 1931: production (Mexico–gas)	One-country experience; only production of gas
Murphy	no involvement abroad	
Occidental	no involvement abroad	
Phillips	no involvement abroad	
Pure	1926: exploration (Venezuela)	Strategy of having a portfolio of undeveloped acreage
Richfield	1926: marketing (Mexico) 1930s: export business	Attempt of direct marketing in Mexico; then export only
Signal	no involvement abroad	
Sinclair	1916: exploration (Angola; Mexico) 1919: marketing (Mexico; Cuba) 1920s: exploration (Western Hemisphere) 1920s: production (Mexico) 1930s: refining (Mexico)	Totally vertically integrated strategy abroad: a deliberate policy

Exhibit 2: Continued

Company	Evolution of International Strategy	Emphasis of Strategy
Skelly	no involvement abroad	
St. Indiana	1925: acquires control of Pan American with exploration, production, refining (Mexico; Venezuela), and marketing (world-wide)	Integrated company; sold all foreign interests in 1932 to Standard (New Jersey)
St. Ohio	no involvement abroad	
Sun Oil	1909: marketing (U.K.–lubricating oils) 1930: exploration (Venezuela) 1930: direct marketing (Canada)	Accent on marketing–direct and export exploration in one country and in joint venture with Standard (New Jersey)
Sunray	no involvement abroad	
Superior	no involvement abroad	
Tide Water	1880s: export 1920s: direct marketing (Philippines) 1930: exploration (Mexico) 1934: refining (Japan)	Accent on marketing, although exploration in one country and refining in joint venture
Union Oil of California	1907: marketing (Latin America) 1918: exploration (Mexico) 1920: production (Mexico) 1921: refining (Canada)	"Pacific" strategy, that is, emphasis on investment located on the western side of the Latin American continent (Mexico, Colombia, Vancouver, Australia, New Zealand)

SOURCE: Companies' *Annual Reports.*

Exhibit 3
STRATEGIC GROUPS OBSERVED BEFORE WW II

	Between 1910 and 1940	By 1939-1940
Not involved abroad	Ashland, Getty, Murphy, Phillips, Signal, Skelly, Standard Ohio, Sunray, Superior	Same + Marathon, Standard Indiana
Explorers only	Amerada, Pure Oil	Amerada, Pure Oil, Continental
Explorers & Producers only	Marathon (gas)	
Explorers & Marketers only	Sun Oil	Sun, Union Oil
Explorers, Producers & Marketers only	Cities Service, Continental	Atlantic, Cities Service, Sinclair
Refiners & Marketers only		Tide Water
Marketers only	Richfield (direct-marketer)	Richfield (export)
Totally Integrated	Atlantic, Sinclair, Standard Indiana, Tide Water, Union Oil	Sinclair

third, however, put marketing as the first step. Yet for one case–Richfield–it was an abortive attempt, and in another case–Sun–it was not marketing of refined products in general, but more narrowly marketing of lubricating oils. But even when the companies started with marketing, they went back afterwards to the expected pattern of vertical integration. In all, five out of the twelve independents with a foreign involvement prior to World War II implemented a total integration abroad.

The analysis in terms of strategic groups, summarized in Exhibit 3, allowed for the observation of three main strategic groups based on frequencies: (1) explorers only (two firms); (2) companies integrated in exploration, production and marketing (two firms); and (3) totally vertically integrated companies (five firms). Three other isolated patterns were also noted.

The distinction, made in Exhibit 3, between maximum development of international strategy during the 1920s and 1930s and the situation that existed in 1939-1940 shows the high instability of the independents' international strategies during this thirty-year period. For example, of the five independents that had at one time a policy of total vertical integration abroad, none held it by 1940. Standard of Indiana dramatically dropped out of international business in 1932.[2] Tide Water remained only in refining and marketing and Union Oil only in exploration and marketing. The refining facilities of Atlantic and Sinclair located in Mexico were nationalized in 1938, thus breaking their vertical integration chain.

The Great Depression with the domestic difficulties it brought forced some independents to divest partly or totally from abroad. Standard of Indiana, for example, was not in a financial position to consolidate, at the same time, its integrated operations at home and abroad, so it sold the latter to Standard of New Jersey in 1932. Because of the Depression, Richfield became very conservative in the scope of its operations,[3] and Sun Oil had to pull out of overseas.[4]

The Great Depression coincided also with a sharp increase in domestic production capacity brought by the discovery of big fields in Texas and elsewhere in the U.S., thus providing an additional incentive for purely domestic strategies. Accordingly, none of the independents engaged only in foreign exploration was active by 1940.

In addition, the Mexican nationalization of 1938 and the freeze of assets in the European countries at war beginning in late 1939 were two external events that bent decisively already fragile strategies. Cities Service, Continental, Marathon and Sinclair were expropriated in Mexico; Atlantic, too, was in Mexico and in Europe.

II. The Model Applied to the Post-World War II Period[5]

Two main strategic groups were isolated empirically. The results, summarized in Exhibit 4, show that approximately one out of three independents went abroad exploring and, if successful, producing; they constitute the strategic group I—the "explorers-producers." Two-thirds of the independents integrated downstream to marketing; they form the strategic group II—the "totally vertically integrated" independents.

Because of the operational definition used to form the strategic groups—maximum stage of development reached at any time during the period under study—the notion does not allow for backward changes of strategic groups due to divestment between the year of maximum development and the end of the period (1976). For example, Signal got entirely out of the international petroleum scene in 1974, and Standard of Ohio kept only its share of production in the Iranian Consortium after 1961. Two other companies divested also partly: Getty from retail marketing, and Occidental from all foreign marketing activities. In addition, as noted in Chapter I, Exhibit 4, five companies were merged in the 1960s in the still existing independents, thus stopping any potential

Exhibit 4
STRATEGIC GROUPS OBSERVED AFTER WORLD WAR II

Strategic Group I: Explorers-Producers	Strategic Group II: Totally Integrated
Amerada	Atlantic
Ashland	Cities Service
Pure Oil	Continental
Richfield	Getty
Skelly	Marathon
Standard of Ohio	Murphy
Sunray	Occidental
Superior	Phillips
	Signal
	Sinclair
	Standard of Indiana
	Sun
	Tide Water
	Union Oil of California

SOURCE: Analysis derived from companies' *Annual Reports.*

evolution of their international strategies.

Beyond these significant but isolated changes, I observed, however, a high stability within strategic groups as soon as the final stage of development aimed at by a given independent was reached. Excluding the independents that disappeared as separate entities by merger, the "explorers-producers" were in their group in average by 1956, and the "totally vertically integrated" companies were in their group in average by 1959. In other words, positions in strategic groups were reached from ten to fifteen years in average after the first decision to go exploring abroad was made, and in majority they were kept an average of at least fifteen to twenty more years. To illustrate this point, Exhibit 5 gives, by five-year clusters, the distribution of the independents' entries in each stage of the international petroleum industry. It shows that the bulk of integrations into exploration occurred between 1945-1950, into production between 1950-1960, into refining between 1960-1965, and into marketing between 1955-1965.

In contrast to the pre-World War II period with a greater variety of patterns of international strategy, after the War the independents under survey fell neatly into two main strategic groups: explorers-producers, and totally vertically integrated companies. I expected these two groups to exhibit characteristics with significantly different orders of magnitude. More specifically, I hypothesized that two of the proxies–size and self-sufficiency ratio–that played a role in the decision to enter foreign exploration were also behind the

Exhibit 5
Evolution of Entries Into the Different Stages of Foreign Oil Business

Stage	All period	by: 1950	1955	1960	1965	1970	1975
Exploring	All	Sinclair Ashland Skelly Pure Oil Tide Water Amerada Phillip Stan. Indiana Richfield Atlantic Signal Marathon Sun Oil Stan. Ohio Cities Service Union Oil Superior Getty Sunray	Ibid + Murphy Continental	Ibid	Ibid + Occidental	Ibid +	All
Producing	All	Phillips Skelly Atlantic Sunray Getty	Ibid + Sun Oil Superior Sinclair Union Oil St. Ohio St. Indiana Tide Water Signal Marathon Richfield	Ibid + Amerada Murphy Ashland Pure Oil Cities Service Continental	Ibid	Ibid +	All
Refining	All Except: (Amerada Continental Getty Murphy Sunray Pure Oil Richfield Skelly St. Indiana St. Ohio	Sinclair Union Oil Phillips	Ibid + Cities Service Sun Oil	Ibid + Tide Water Continental Getty	Ibid + Murphy Signal St. Indiana Atlantic Marathon	Ibid + Sunray Occidental	Ibid
Whole-saling	All Except: (Amerada Continental Getty Murphy Sunray Pure Oil Richfield Skelly St. Indiana St. Ohio	Tide Water Union Oil	Ibid + Cities Service Sun Oil Superior Atlantic	Ibid + Phillips Signal	Ibid + Sinclair Marathon	Ibid + Ashland	Ibid + Occidental
Retailing	All Except: (Amerada Ashland Pure Oil Richfield Skelly St. Ohio Sunray Superior)	Tide Water	Ibid + Cities Service Sun Oil Atlantic Phillips	Ibid + Continental Murphy Signal	Ibid + Getty Sinclair St. Indiana Marathon	Ibid + Occidental	Ibid + Union Oil

split of the group of independents into totally-integrated and partly-integrated firms. However, in that case, these two proxies stood for slightly different variables.

Whereas in 1946 size stood for the lumpiness of the first investment into foreign explorations, in 1959 (the year in which a wave of entry into foreign refining began) it stood instead for the lumpiness of the first investment in a refinery abroad. I therefore expected the small independent to face what was to it a higher barrier of entry into foreign refining than was the case for a big firm.

Also, whereas in 1946 a crude short independent (that is with a self-sufficiency ratio lower than 1) had a greater incentive compared to a crude long firm to go exploring abroad because of its need to feed its domestic refineries, the situation was reversed when production was developed abroad. I expected a crude long world-wide independent to have enough foreign production to consider locating a refinery abroad as a more efficient use of its resources and thus as a profitable investment. On the contrary, I expected a crude short independent to keep its initial objective of importing back in the United States its foreign production in order to feed its domestic refineries, and not to risk duplicating abroad its unbalanced domestic vertical integration chain.

As shown in Exhibit 6, in terms of relative size in 1959 the companies that were to integrate later were on the average more than twice the size of the

Exhibit 6
SIZE AND CRUDE BALANCE DIFFERENCES BETWEEN THE TWO STRATEGIC GROUPS

	Firm's Size (in % of Exxon)		Firm's World-Wide Self-Sufficiency Ratio (own production/ refinery runs)
	In 1959	In 1945	
explorers-producers	3.9	4.1	.52
totally vertically integrated firms	8.5	15.2	.81

SOURCE: Data derived from companies' *Annual Reports.*

companies that did not. A test showed that the difference was significantly below the 5% level. The same test repeated on the sizes in 1945 was significantly below 1%. Self-sufficiency ratios were also significantly different, although less from a statistical point of view. In average the nonintegrated independents were substantially more crude short than the integrated ones in 1959. Moreover, as shown in Exhibit 7, the small independents that integrated totally later on were generally crude long in 1959, whereas the small ones that stopped at foreign production were generally crude short. On the contrary, all the big independents, except Pure, integrated totally abroad.

Exhibit 7
CROSS-TABULATION FOR SIZE AND SELF-SUFFICIENCY FOR THE TWO STRATEGIC GROUPS*

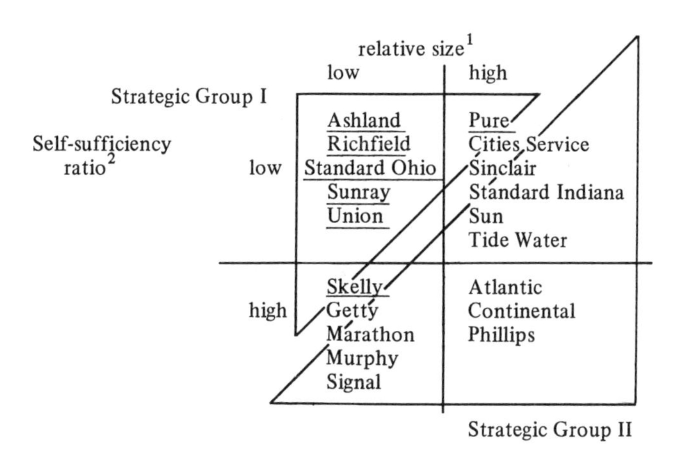

* 1959 data.
[1] The cutoff point between "high" and "low" for size is the average size of 6.8% of Exxon's.
[2] The cutoff point for self-suffiency (production/refinery runs) is .5.
The names underlined are those of the non-integrated independents involved only in exploration and production.

III. Influence of Pre-World War II International Experience on Post-World War II International Strategy

If having pre-World War II experience abroad or not seems to have had no influence on the decision of entry into foreign exploration,[6] it seems to have strongly influenced the patterns of international strategy adopted by the independents later on.

Of the twelve independents involved abroad before World War II, four were found in the strategic group I and nine in the strategic group II. Of the nine independents[7] not involved abroad before World War II, five were found in the strategic group I and four in the strategic group II.

From this cross-tabulation, I concluded that a low foreign involvement profile after World War II was linked with a similar behavior before World War II. On the contrary, an extensive involvement abroad before World War II was followed by a strategy of total vertical integration abroad in the more recent period.

NOTES

[1] See Appendix 4 for a detailed account of the independents' activities abroad before World War II.

[2] See Appendix 4 for further details.

[3] Information obtained during interview in Atlantic-Richfield.

[4] Information obtained during interview in Sun.

[5] Because the characteristics of each group are described in the two following chapters, this section will give only an overview on the results in order to help the reader position both strategic groups and companies in each group.

[6] See Chapter 3, Section V.

[7] Not counting Occidental who began to be an actor in the oil industry only in the 1960s.

CHAPTER VI

STRATEGIC GROUP I: EXPLORERS-PRODUCERS

Introduction

The present chapter analyzes the first strategic group, whose members went abroad only to explore and produce, that is, who did not integrate forward into refining and marketing abroad. In a second section, I present a detailed case study whose purpose is to show that what may look like a neat strategy ex post, was in fact a lengthy formulation-implementation process.

I. Comparative Analysis of the Strategic Group's Members

This section is devoted to an examination of the entire strategic group of the explorers-producers. Differences and similarities in patterns of exploration over time are pointed out. Responses to changes in the external environment are evaluated.

Impact of the 1959 Import Controls. Because all the members of this strategic group, except one–Skelly–began producing crude oil outside the North American continent around 1959, as shown in Exhibit 1, I hypothesized that the mandatory import controls imposed by the U.S. government in March 1959 were a major disturbance in the firms' international policies. Though two were just domestic producers–Amerada and Superior–the others were generally badly crude short: four out of six had a self-sufficiency ratio of about 50% or less the year preceding the beginning of their production outside the U.S. and Canada (See Exhibit 2).

Because not involved in refining and therefore without import quota allocation, Superior seems to have been the hardest hit, judging from its reaction. Realizing that its full production was not being utilized, particularly because of mandatory import controls, Superior's management endeavored to effect a merger with Texaco in 1959. But opposition from the Anti-Trust Division of the Department of Justice forced abandonment of this solution.[1] The Venezuelan operations, which were jeopardized by the controls, were eventually sold to Texaco in January 1964 for Texaco stocks.[2] This setback did not stop Superior from exploring abroad, although its explorations were rather unsuccessful except in the North Sea (U.K.) where it was associated with Phillips.[3]

Although the initial import quota allocated to Standard of Ohio (Sohio) roughly matched its foreign production, it stopped the acquisition of new concessions abroad after 1960. In a step further, in 1963, it signed long-term

Exhibit 1
Strategic Group I: Evolution of International Strategies

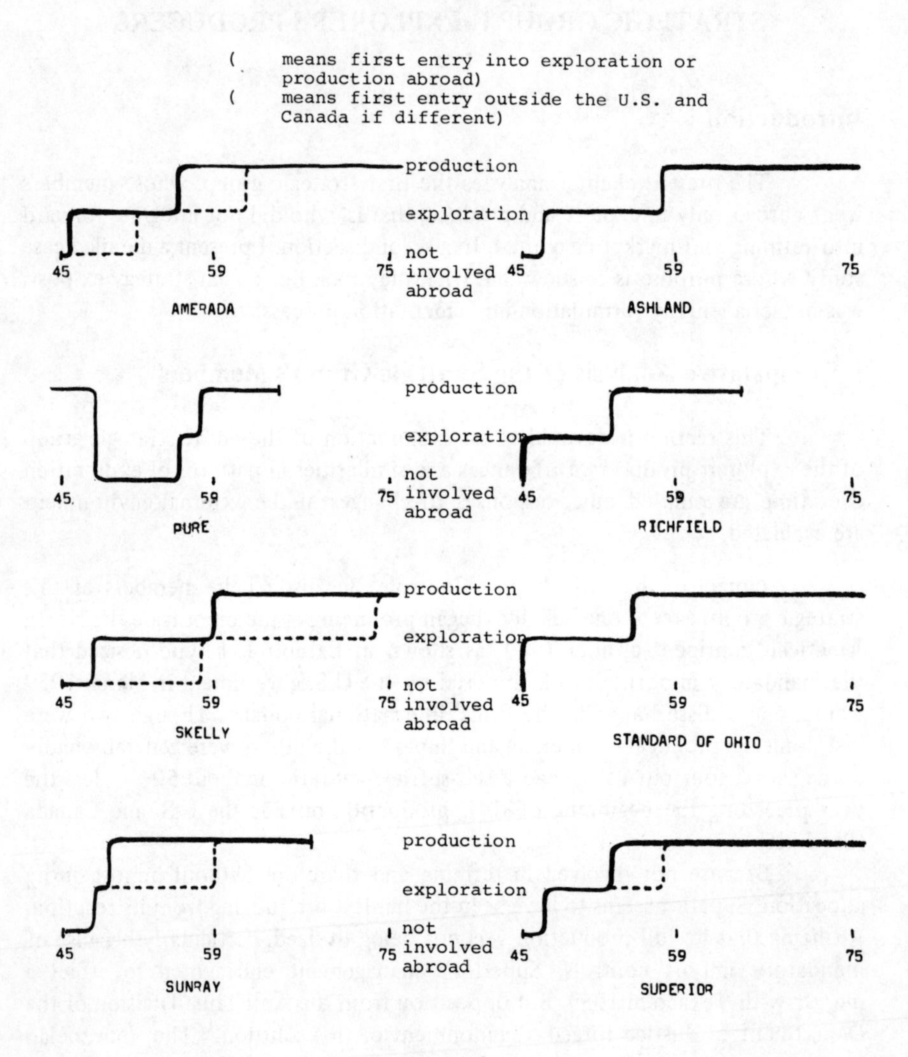

Exhibit 2
SELECTED STATISTICS ON SOME INDEPENDENTS' EARLY PRODUCTION ABROAD

Company	Date of First Production Outside the U.S. and Canada [a]	Self-Sufficiency Ratio (Own Production/ Refinery Runs) in the Year Preceding [a] [b]	Foreign Production in 1960 in b/d [c]	Import Quota in 1960 in b/d [d]	[d] in % of Quantities Refined in the U.S. the same year [e]
Amerada	1962	p[1]	5,625[2]	n.a.[3]	--
Ashland	1958	6.3	15,300	13,275	7.5
Pure Oil	1958	44.5	12,176	14,762	8.4
Richfield	1959	50.6	12,582	24,551	18.0
Skelly	1974	98.3	1,896[2]	5,380	--
Standard Ohio	1958	29.9	11,302	12,707	7.9
Sunray	1959	72.1	4,285	13,826	9.0
Superior	1958	p[1]	28,119	n.a.[3]	--

SOURCE: Companies' *Annual Reports* and *Oil and Gas Journal*, December 21, 1959, p. 54.
[1] p = production only. [2] Canada only. [3] n.a. = no allocation.

contracts for the purchase of Middle Eastern and Venezuelan crude to fill all its foreseeable needs. As the company management justified the reversal of strategy for its stockholders: "This will enable us to share in the essential advantages of foreign crude imports without incurring the risks of investment in overseas exploration and production."[4] Sohio kept its foreign production, however, in the Iranian Consortium and in Venezuela. In January 1965, it bought a majority interest in a Canadian company with a small production in application of a new policy formulated the year before of acquisition and development of existing oil reserves.

Pure was the keenest to protest against the unfairness of the import restrictions: "One of the greatest obstacles facing oil companies today is the difficulty of planning future operations with respect to import of crude. . . . We feel strongly that the plan now in effect favors historical importers and does not recognize sufficiently the position of those companies, such as Pure, which have just established foreign production of crude and have the opportunity to sell the resulting products in their normal marketing areas."[5] In 1961, Pure complained that its import quota had been reduced by 2%, but its foreign production that year was 16% below the U.S. Government allocation. Furthermore, in terms of the quantities that Pure refined in the United States, the import quota was as low as for the other refiners–with one exception. (See Exhibit 1.)

Ashland, both badly crude short and a small-foreign producer, was more ambiguous about the administration of imports. One year, when discovery had just been made in Venezuela, it regarded the allocation of import quota as an allocation of domestic markets because of the price advantage of foreign crudes over domestic ones. But the following year, when it imported the double of its share of Lake Maracaibo's production, it was quick to point out the advantages of import quotas from a profit point of view: "We are realizing approximately one dollar per barrel more for this oil [from their concession in Lake Maracaibo (Venezuela)] than if it had to be sold on the open foreign market without an accompanying import quota."[6]

Sunray and Richfield did not raise any complaints about the import controls in their annual reports, though both were merely half self-sufficient.[7] But Richfield was awarded an import quota twice the size of those allocated to its fellow refiners (see Exhibit 2). Sunray, with an initial quota equal to three times its foreign production, quickly merged in October 1959 with Suntide Refining Company, a 55,000 b/d refiner in Texas, in order to provide an outlet for its Venezuelan crude oil and to benefit from the refiner's import quota allocation.[8]

As hypothesized, the coincidence of new and growing foreign productions and of tightening import controls caused these independents either to revise their international strategies or to limit the scope of their involvement abroad to what they had done so far. It forced Sohio to divest almost entirely from abroad. It probably also later dissuaded Amerada from importing its Libyan crude oil to the United States.[9] Yet, the fact that seven out of eight independents in this strategic group kept on implementing the same international strategy of exploration and production supports the conclusion of high stability of strategy over time, that is, through obstacles represented by changes in the external environment.

Intensity of the Exploration Effort Over Time. Although all the independents in this group, when active, acquired between one and two new concessions per year, a notable diversity of intensity of exploration efforts over time was observed. The intensity of exploration effort was defined as the number of countries or oil areas entered by a firm between 1945 and 1976 normalized by the number of years of active exploration. This ratio ranges from less than one entry every three years (Sohio) to almost two each year (Pure); the average and the median being around one new country/concession entered/acquired per year.

These differences allowed regrouping the independents as either "continuous" explorers, or "cyclical" explorers. The results are summarized in Exhibit 3. The four explorers with a continuous stream of new additions to their portfolios of foreign concessions, however, made short pauses (generally one year) after discoveries in order to develop them. On the contrary, cyclical explorers, whose efforts were concentrated on separate sub-periods, had bursts of up to two concessions acquired each year when active.

Exhibit 3
INTENSITY OF EXPLORATION ABROAD OVER TIME

Pattern of Exploration Effort	Name of Independent	Number of Concessions Acquired Per Year
continuous	Pure	1.8
	Superior	1.3
	Richfield	1.25
	Amerada	1.2
cyclical	Ashland	0.8
	Sunray	0.7
	Skelly	0.45
	Sohio	0.3

SOURCE: Companies' *Annual Reports.*

A correlation test showed that the intensity of exploration was significantly associated with the company's crude position (measured in 1959): the higher the firm's self-sufficiency ratio, the lower the intensity of its exploration effort over time, and conversely.[10] In other words, the crude short independents were the most prone to explore actively abroad in order to attain a better balance of their vertical integration chain.

I found, however, no significant relationship between the firm's size and its intensity of exploration abroad. It seems that, after the first entry, the foreign exploration no longer presented a barrier to the small independents.

Geographical Patterns of Exploration Over Time. The explorers followed two distinct geographical patterns in their acquisitions of concessions abroad. Four of them spread their efforts worldwide throughout the thirty-year period whereas three focused first almost exclusively on the Western Hemisphere and then switched as exclusively to the Eastern Hemisphere when import controls were set up in 1959. Sohio started with a Western Hemisphere exploration strategy, but stopped exploring abroad after 1960.

Exhibit 4 shows evidence of a relationship between the level of self-sufficiency and the geographical pattern adopted. Crude short independents went exploring first within close range from a transportation point of view, having in mind to bring any production back into the United States. When

Exhibit 4
GEOGRAPHICAL PATTERNS OF EXPLORATION

Company	Self-Sufficiency Ratio[1]: The Year of Entry Abroad	Self-Sufficiency Ratio[1]: In 1959
Worldwide Exploration Strategy		
Skelly	1.84	1.31
Sunray	1.24	0.48
Amerada	P[2]	P
Superior	P	P
Western-then-Eastern Hemisphere Strategy		
Ashland	0.22	0.15
Pure	0.44	0.36
Richfield	0.37	0.44

[1] Own Production/Refinery Runs: A t test for two means (taking the average of each pair of figures) showed that the two exploration strategies were associated with different self-sufficiency ratios (test signficant below the .5% level).

[2] P = only a producer; companies in that case were not included in the test.

import quotas were no longer advantageous, they just went where the potential oil areas were. "In its continuing effort to increase foreign production, Pure for the first time expanded its search for oil to include the Eastern Hemisphere."[11] The independents with no pressing need to feed their refineries or with no refinery at all, on the other hand, aimed directly at the potential areas from a geographical standpoint regardless of the geographical location. Superior, for example, has had always the policy to explore where the best places are.[12]

Joint Ventures in Foreign Exploration. Because of the risks involved in the exploration phase of the oil industry, the joint venture is commonly used. With the additional risks implied by international exploration, especially from the point of view of small firms, joint venturing was an obligation for the independents immediately after World War II. Skelly, for example, went abroad through joint ventures because of the company's size.[13] In 1956, Pure stated clearly that: "In order to spread our risks and to provide a much wider participation in favorable areas, we have elected to join forces with other American oil companies with interests somewhat similar to our own."[14] It was clearly advised after the end of World War II that "more of the smaller independent American companies pool their resources for foreign undertakings."[15]

They eventually went beyond that, going through the first stages of the learning curve with "institutionalized" joint ventures. As shown in Exhibit 5, five out of the eight explorers in the strategic group went practicing the foreign oil business in either especially set corporations or with the same partners repeatedly. Furthermore, these special-purpose joint ventures were with other independents who, in all cases, happened to implement a strategy of vertical integration abroad. It is no mere coincidence, but there is no evidence that those latter independents knew in the late 1940s that they would integrate operations abroad as they did at home. At the time they entered the international scene, all the independents were just explorers.

Exhibit 5
FIRST ENTRIES ABROAD THROUGH JOINT VENTURES

Amerada	Conorada (Continental, Marathon) (Between 1948-1962)
Ashland	Aminoil (Phillips) (Between 1947-1970); Amurex (Murphy) (Between 1951-1958)
Richfield	Sinclair, Cities Service (Between 1945-1959)
Skelly	Getty (Between 1948-1976)
Sunray	Phillips (1956-1963)
St. Ohio	Joint ventures with various partners
Superior	Joint ventures with various partners
Pure –	Joint ventures with various partners

SOURCE: Companies' *Annual Reports.*

Reactions to the 1973 Oil Crisis. Although only four explorers were still involved abroad by the end of 1976—three among the four others being merged in the meantime and the fourth, Sohio, having stopped exploring fifteen years ago—it is possible to draw some conclusions about their reactions to the oil crisis of 1973.

As shown in Exhibit 6, three out of the four reduced considerably their acquisitions of foreign concessions between 1974 and 1976, compared with 1971 to 1973. They also tended to avoid completely the O.P.E.C. countries. During this period, Ashland, for example, was pulling out as quickly as possible from O.P.E.C. areas.[16] Still the large majority of the independents' efforts were directed toward less developed countries potentially less safe than developed ones.

Exhibit 6
EXPLORATION PATTERNS ABROAD BEFORE AND AFTER THE 1973 OIL CRISIS

	Concessions Acquired in Period 1974-1976			Total Concessions Acquired in	
		Less Developed Countries			
	Industrialized Countries [a]	All LDCs Including OPEC [b]	(Among which Non-OPEC Countries)	1974-1976[1]	1971-1973
Amerada-Hess	1	3	[3]	4	7
Ashland	1	2	[1]	3	6
Skelly	1	1	[1]	2	2
Superior[2]	1	8	[7]	9	14

SOURCE: Companies' *Annual Reports.*
[1] [a] + [b].
[2] Including acquisitions by the Canadian subsidiary.

II. A Case Study: Amerada/Amerada-Hess

The objective of this section is to describe in great detail the international strategy of one company belonging to the first strategic group.

From its entry abroad until 1969, the date of its merger with Hess, Amerada's international strategy has been unchanged and limited to exploration and production. It was the strategy followed at home, and it was also the strategy implemented before World War II. For fifty years, the policy was to be involved only in exploration and production, and that choice guided the extent of foreign operations as well. By the end of the War, Amerada had the reputation of the company mainly responsible for developing geophysical research in the U.S. This reputation was out of proportion to its size. Its 1945 gross income amounted to $27.2 million, which puts it "in the class of small-to-middling corporations."[17] By comparison with the majors, it was minuscule. But, as *Fortune* put it, "of crude-oil producers that have no refining facilities and no pipelines, Amerada was the biggest." As President of Amerada since 1929, Jacobson had been behind the formulation of three safe-and-sound company policies: (1) to stay in the crude business and not mix in transportation, refining, or marketing; (2) to be ultra-conservative financially; (3) to keep up aggressive exploration to maintain oil reserves. On this last item, by 1945, Amerada had been more aggressive than most, with one reservation–it has stayed mainly within the U.S.

The Reentry Abroad. Amerada was present abroad before the war, but not significantly. The foreign portfolio of concessions consisted of a control interest acquired in 1926 in a company named Esperenza Petroleum, which had large concessions in Venezuela. The company deliberately made no effort to explore the promising acreage, although it spent cumulatively large amounts of money just to keep the concessions. This discrepancy led to a situation where Amerada reported on its Venezuelan holding in terms of information coming from competitors. For example, when the petroleum laws of Venezuela were modified in 1943 so as to increase considerably the cost of carrying undeveloped acreage, the company justified its unchanged position to its shareholders in the following way: "Active exploration by a number of other companies is now in prospect and may help to evaluate some of the properties and be of assistance in formulating plans for their development or other disposition."[18] Amerada eventually kept its interest, but ended by paying more taxes to the Venezuelan government because it did not change its policy of undeveloped acreage. For example, while advancing $158,000 to Esperenza, it acknowledged that "during 1947 a successful test well was completed by another company approximately three miles from a group of concessions held by Esperenza in the State of Barinas, but until additional wells are drilled, the importance of this discovery in relation to the Esperenza properties cannot be determined. . . . Pending further developments in this immediate area, no plans are being considered for drilling on Esperenza's holdings."[19] But Amerada's policy was not revised when a second producer was completed the following year. "Apparently an oil field of as yet unknown extent and importance has been discovered" was the company's only comment.[20]

The idle tactics chosen in the Venezuelan venture contrast with the aggressive strategy implemented elsewhere at the same time. In the late part of

1947, Amerada extended its operations into the Province of Alberta (Canada). To the end of March 1948, large lease options, as well as exploration rights with selection options known as "reservations," had been obtained on 1,150,000 acres of land; nine months later the acreage had doubled. This time, geological and geophysical exploration work had been commenced immediately and were "expanded as rapidly as possible," as announced.[21] Amerada, in fact, was one among a number of companies that entered the scramble for acreage following the discovery of Leduc field by Imperial in the Spring of 1947.

Despite the precedent of the Canadian and Venezuelan operations, Amerada's first entry into the international petroleum scene is often associated with the formation of Conorada (a name that stands for CONtinental, Ohio, and AmeRADA) near the end of 1948. The same mistake applies also to Amerada's two other partners, "none of which had engaged in foreign operations in the past," as witnesses the *Oil and Gas Journal's* comment on the formation of Conorada.[22] Judging by the importance given to the move in each company's annual report, it apparently meant the most for Amerada. Whereas Ohio Oil (later renamed Marathon) qualified the venture of "potentially large scope" but not designed to "undertake drilling and production operations,"[23] and Continental waited until its 1949 *Annual Report* to release the news, Amerada described the deal in every detail: "The purpose of the new corporation is to investigate the possibilities and prospects for the development of oil production in foreign countries, other than Canada [where Amerada and Continental were already]. Conorada has only a nominal capitalization, and necessary advances will be made by the three parent companies from time to time as required. The arrangement among the three companies is very flexible, and each has reserved the right to decide in respect of each proposal that may be offered, whether or not it desires to participate. If any foreign properties are acquired, it is intended that the acquisition and subsequent development of such properties will be carried on either directly by the interested parent companies or by one or more subsidiary companies which may be organized for such purposes."[24]

The first approach to international petroleum business through joint venture is typical of the majority of the independents, though the forms differed. Conorada was a sort of "staff" joint venture for the study of the foreign oil environment, not an operating company. This is witnessed by the choice of an internationally known consultant in geology and petroleum engineering for the position of general manager; this consultant's firm had advised oil-producing countries in Latin America and the Middle East on the drawing of petroleum laws.[25]

In 1949 the first corrections were made on Amerada's new international strategy. Not only did Amerada choose to go through a joint venture after a trial alone in Venezuela and Canada, but after eighteen months of unsuccessful exploration in Alberta, Amerada moved to the direct purchase of proven oil land. Although it was not an entry into production without prior exploration—which Amerada never did—it was no mere wildcatting either. In June 1949, Amerada purchased 640 acres of proven oil land for $3.1 million from the government of Alberta. Drilling was commenced immediately, and by the end of October sixteen wells, qualified as "good producers," had been completed. Amerada had to wait until 1952 to find oil (700 b/d) in Alberta as a direct result of its wildcatting activity.

It took as long for Conorada (1949-1952) to go from observation of the international oil environment to entry into actual exploration abroad. In 1952, through the joint venture, a one-sixth interest was acquired in a concession covering 90,000 square miles in Italian Somaliland. The same year, a company formed for the occasion, Conoco Somal Ltd., acquired 100% interest in 36,000 square miles in British Somaliland. These tentative explorations at the limit of the Middle East were unsuccessful, and the concessions were relinquished in 1966.

In 1953, after new discoveries in Alberta, the company undertook a preliminary study of a 120-mile pipeline to Vancouver. It was a departure from a long-set policy of not being involved in transportation. But this change of strategy was dictated by the necessity to find an outlet to the Canadian crude oil. A company, in which Amerada took "a small stock interest [11%]," [26] was formed in 1954 to build the projected pipeline. Shipments commenced in March 1956. Amerada had gone from Stage 1—exploration—to Stage 2—production—in nearly nine years. But, as told earlier, *it never went beyond.*

A posteriori, Amerada's entry into Libya may look, at first glance, like a turning point in its international strategy. But Libyan discoveries did not change Amerada's scope of operation as it did for its two other partners in Conorada. When, late in 1955, Amerada acquired a one-third interest in a concession covering 55,000 square miles, it was merely a new exploration venture, though a well-prepared one, since geological work had been carried on for two years by Conorada. Yet at the time it was nothing more than another gamble.

This feature of the exploration stage is particularly enhanced when the entry in Libya is paralleled with the decision not to reenter Venezuela, another oil area—but one already proven. When, in January 1956, Venezuela announced that applications for concessions would be entertained, applications made by Amerada in prior years (jointly with others) were confirmed or modified.[27] But in the end no acreage was acquired by the company. The explanation provided by the management was that: "The Venezuelan Government granted at very high prices a few concessions of limited area principally in the Lake Maracaibo District."[28] However, as in Libya, Amerada's intentions were clearly indicated as a limited amount of seismic work was conducted in 1955 and even expanded in 1956.

Yet this new attempt in Venezuela contrasts with the conservative behavior in the then 30-year-old Esperenza venture. While Amerada was complaining about high bidding prices, it paid the same year $247,000 to the Venezuelan government just to keep Esperenza's concessions idle, although they were promising by all indications. For example, as for almost every year since 1938, Amerada noted the progress of its neighbor competitors in Venezuela the following way: "Two other companies which have developed production closest to some of Esperenza's holding in the Barinas area are presently constructing a 200-mile, 20-inch pipeline with an initial capacity of 100,000 b/d."[29] Still Esperenza did not participate in the bidding opened in February 1957 in the Barinas District.

In the late 1950s, besides an entry into Australia, the main new development of Amerada's international strategy was the beginning of a policy formulation about the Libyan oil. The first discovery was made in 1958, and the 1959 *Annual Report* pointed out its strategic importance: "While Libyan production will be coming into a world market already fully supplied with oil, it will, due to its geographical location, be in an advantageous position with respect to cost of transportation to the principal European and North American markets." Amerada, however, was not to exploit these market opportunities. In March 1962, the company entered into an agreement for the sale of its share of production from one concession to its partner Continental. The deal, which cost Amerada 7.5 cents per barrel deducted by Continental as a marketing fee, lasted until June 1965.[30]

In the meantime, Amerada was consolidating even more its policy—new abroad, but old at home—of selling its crude oil at the well. In 1962 Amerada started discussions with Royal Dutch-Shell for the sale over a long period of time of substantial quantities of Libyan oil. The negotiations lasted about two years. According to Amerada's management, such delay was required "in resolving the many problems necessarily involved in concluding a transaction of this nature."[31] "The working out of a transaction of this nature and scope is of necessity very complicated."[32] Negotiations with Shell were virtually completed by mid-1964, but approval by the Libyan government waited until January 1966.

The reasons for "one of the longest commercial courtships on record"[33] were obvious to both parties. "The principal advantage of the Shell agreement to Amerada is that it will enable Amerada, without any investment in very costly refining and distributing facilities, to benefit from the marketing of products in the natural market for Libyan oil—which is at present Western Europe—and to obtain a profit geared to the realization obtained in this market by one of the world's largest, long-established and most efficient refining and marketing organizations."[34]

This move secured for Shell a source of cheap crude near Europe—as near to this vital market as its rivals BP and Exxon, also in Libya. It was also a step in Shell's struggle with its own chronic shortage of crude.

Practically it was a transfer of parts of concessions, which explains why approval of the Libyan government was required. To obtain joint-production with Amerada, Shell reimbursed the American company "of about $64 million corresponding to approximately half of Amerada's expenditures in Libya from the beginning of operation in 1955 to January 1, 1966."[35] The 1964 agreement provided that Shell would take its own plus Amerada's share of the joint-production up to 270,000 b/d. Since August 1963, all the production not covered by the Conoco contract has been sold to Shell under an interim agreement.

These two successive deals solved all the marketing problems of Amerada's first production in the Eastern Hemisphere. It seems that there was no attempt to integrate downstream by taking the opportunity to go into refining and marketing in Europe. Such an exclusion was consistent with the policy formulated some twenty years earlier and with the fact that it was still the same top management to implement it. A. Jacobsen was President in 1929 and E. H. McCollough was mentioned as Vice President back in the 1940s.[36] In 1955, when Amerada was on the verge of entering Libya, Jacobsen was named Chairman and McCollough, President. In 1965, when the Conoco agreement expired and the Shell deal was already working through an interim agreement, McCollough was elected to the additional post of Chief Executive, with Jacobsen still Chairman. It is, therefore, not surprising that the stability in top management went along with a continuity of policy over time.

Conservatism did not exclude error, however. Operations in Venezuela are typical. Near the end of 1960, Esperenza farmed out part of its acreage to a major oil company, Mobil. Amerada's subsidiary retained a small interest but was relieved of further tax payments. Less than one year later the farmer made a discovery with an initial flow of 5,000 b/d. A second producing well was completed early in 1962. The irony of the Venezuelan venture is that, after 34 years, and $4.5 million spent in taxes, and in spite of encouraging signals given by discoveries made nearby, Amerada never decided to drill on its holdings.

But as soon as it was done—but by another operator—it was successful. Certainly it was small compared to the Libyan bonanza and it is no coincidence that Amerada decided to give up in Venezuela at the time when development started in Libya. Yet, had it decided to get into active exploration on its own and sooner, Amerada would have gotten a production comparable to its Canadian net interest. Instead, in 1967 its 68% interest in Esperenza was still carried on, at cost less reserve, for $1,114,000.

Once the development of production in Libya was completed, some restructurings of Amerada's exploration efforts took place. The first step was to dissolve Conorada, the joint venture used to get into exploration outside North America. Because of diverging interests among the three partners, Amerada bought up Conorada from Continental and Marathon in 1963.[37] Two years later, the staff and facilities of this wholly owned subsidiary were integrated in Amerada's international organization. *When the company's maturity in international business came, the original vehicle used to share the learning costs of doing business abroad was no longer needed.*

The second step in restructuring its exploration portfolio was for Amerada to relinquish concessions in various continents while at the same time investing in potential new oil areas. For example, the first concession acquired through Conorada in Somali was relinquished in 1966 after fourteen years of unsuccessful effort. In 1967 a concession in Tunisia was surrendered after ten years and one dry hole.

In Australia since 1959, Amerada started to increase its participation from 33% to 44% in 1963, to 61% in 1964 and 70% in 1965. In 1967 some acreage was surrendered, and the following year, in absence of results, participation in Australia was reset at 33%.

The same year it increased its gamble in Australia, Amerada participated with three other companies in a reconnaissance service program over a portion of the North Sea. While no concession rights were then available, the work was done in anticipation of such an award.[38] In effect, in 1965 Amerada was granted interests in five concessions by the British government and in three by the Norwegian authorities. In 1966, gas was discovered on the United Kingdom portion of the North Sea and sold to the British Gas Council, one partner in the joint venture.

All these efforts around the middle sixties were part of a deliberate company policy: "The search for new sources of petroleum is being vigorously maintained," explained Amerada's management in the 1966 *Annual Report.*

A simple glance at data around 1969 would not show any radical change between Amerada's international strategy and that of Amerada-Hess. But the 1969 merger being the addition of a domestic refiner and marketer—Hess Oil and Chemical Corporation—on top of an international explorer-producer—Amerada—changed the relative weight of the international strategy in the overall strategy. Hess taking the control over Amerada, the objective was no more to be a successful international explorer but an integrated company balanced between foreign production and domestic refining and marketing activities.

This change of scope altered the role attributed to foreign exploration, with an accent on quicker productivity than before. That led to two modifications in the implemented policy. Firstly, foreign exploration was intensified: as many new concessions were bought in the nine years between 1968-76 as in the twenty-one years between 1947-67. Secondly, the portfolio turnover was

Exhibit 7

STRATEGIC GROUP I – PATTERNS AND THEIR CAUSES (1945-1976)

Causes \ Patterns	Degree of Vertical Integration Limited To Exploration-Production	Intensity Of Exploration Over Time	Geographical Distribution Of Exploration Over Time
1959 Import Controls	Triggered	Increased	Shift to Eastern Hemisphere for Companies With Western Strategy
Size	Influenced: Small Firms	No Influence	No Influence
Self-Sufficiency Ratio	Influenced: Crude Short Firms	Lower the Ratio, Higher the Intensity	Crude Short: Western-then-Eastern Hemisphere Strategy Crude Sufficient: Worldwide Strategy
1973 Oil Crisis	No Influence	Decreased	Avoid OPEC Countries, But Not Other LDCs

SOURCE: Section 1.

sharply increased, that is, concessions were kept a shorter time before relinquishment than before: an average of two years compared to five-to-ten times longer before 1968.

To feed the domestic refining facilities–in particular, the large St. Croix refinery on the U.S. Virgin Islands, dependent on imports for more than three-fourths of its crude supplies–foreign production was from then on either imported back directly or swapped for better-suited or better-located crudes. For example, the Virgin Islands refinery does not need low sulphur crude. So Amerada-Hess can sell its Libyan production and buy cheaper high sulphur crude elsewhere.

The 1973 oil crisis led Amerada-Hess to increase the emphasis on U.S. activities, in particular in Alaska. Abroad, efforts were focused on evaluation and development of properties previously acquired, such as in the North Sea and offshore Iran and Abu Dhabi, where by the end of 1976 it was busy producing 18,000 b/d and developing as much. The few countries where Amerada-Hess acquired new concessions between 1974 and 1976 were significantly located outside the O.P.E.C. cartel, but two of them, Angola and Vietnam, happened to be even worse political bets.

Conclusion

As a conclusion, the analysis made in Section 1 is summarized in Exhibit 7, with an emphasis on causes-and-effects relationships.

NOTES

[1] See Superior, *Annual Report 1959.*

[2] This information was obtained during an interview in Superior.

[3] See Superior, *Annual Report 1966.*

[4] Standard of Ohio, *Annual Report 1963.*

[5] Pure, *Annual Report 1958.*

[6] Ashland, *Annual Report 1959.*

[7] Data derived from Richfield's and Sunray's *Annual Report 1959.*

[8] See Sunray, *Annual Report 1959.*

[9] See section II for a detailed account of Amerada's international strategy since 1945.

[10] The linear correlation coefficient was -.24, with a t test significant below 5%. When the company was a producer, it was excluded from the test.

[11] Pure, *Annual Report 1959.*

[12] Information obtained during an interview in Superior.

[13] Information obtained during an interview in Skelly.

[14] Pure, *Annual Report 1956.*

[15] *Oil and Gas Journal,* "Design for Petroleum Expansion Outside the United States," June 15, 1946, p. 90.

[16] Information obtained during interviews in Ashland.

[17] "Amerada plays them close to the chest," *Fortune,* January 1946. The rest of the paragraph is also derived from information contained in this article.

[18] Amerada, *Annual Report 1943.*

[19] Amerada, *Annual Report 1947.*

[20] Amerada, *Annual Report 1948.*

[21] Amerada, *Annual Report 1947.*

[22] *Oil and Gas Journal,* January 13, 1949, p. 46.

[23] Ohio Oil, *Annual Report 1948.*

[24] Amerada, *Annual Report 1948.*

[25] *Oil and Gas Journal,* January 13, 1949, p. 46.

[26] Amerada, *Annual Report 1954.*

[27] Amerada, *Annual Report 1955.*

[28] Amerada, *Annual Report 1956.*

[29] Ibid.

[30] The 7.5 cents figure was provided in an interview with Leon Hess, Chairman.

[31] Amerada, *Annual Report 1962.*

[32] Amerada, *Annual Report 1963.*

[33] *The Economist,* January 22, 1966.

[34] Amerada, *Annual Report 1965.*

[35] Amerada, *Annual Report 1966.*

[36] *Fortune,* op. cit., January 1946.

[37] This information was obtained during interviews in Continental.

[38] Amerada, *Annual Report 1963.*

CHAPTER VII

STRATEGIC GROUP II:
TOTALLY VERTICALLY INTEGRATED INDEPENDENTS

Introduction

In this chapter, I examine the strategies of those fourteen U.S. independents who chose to get involved abroad beyond the exploration and production stages to the point of being internationally integrated as they are domestically.

A first section is devoted to a comparative analysis of the independents' international strategies. In a second section, a case study–Atlantic/Atlantic-Richfield–is presented as an illustration of the complexity of the process of multinationalization met by the independents.

I. Comparative Analysis of the Totally Integrated Independents' International Strategies Between 1945-1976

The format of this section is to start with foreign exploration, then to go to production, refining, and marketing.

The foreign exploration policies of the group's members may be compared in terms of both the stability over time of their geographical distribution and the intensity of their efforts through the period.

Geographical Patterns of Exploration Over Time. All the fourteen integrated independents followed geographical patterns of exploration abroad between 1945 and 1976 that can be regrouped in two categories as shown in Exhibit 1. The majority of the companies–nine out of the fourteen–started by focusing their acquisitions of concessions in the Western hemisphere with few or no entries elsewhere. Then, by 1960 in average, they all switched their interest to the Eastern hemisphere, with few entries or reentries in Canada or Latin America. The switch coincided with the imposition of mandatory import restrictions in the U.S. This change in their environment prevented the continuation of an exploration strategy relying solely on the premise that all discovered crudes will be exported to the U.S. Since there was no more logistic incentive to confine themselves to the Western hemisphere, they moved outside. Yet they did not do so casually, but as a total commitment, buying new concessions practically exclusively in the Eastern hemisphere from then on.

A smaller group of independents–five out of the fourteen–distributed

Exhibit 1
GEOGRAPHICAL PATTERNS OVER TIME OF EXPLORATION POLICIES

Foreign Exploration Emphasizing Western Hemisphere Then Eastern Hemisphere	Foreign Exploration About Equally Distributed Between Two Hemispheres Through Whole Period
Atlantic	Continental
Cities Service	Getty
Murphy	Marathon
Phillips	Occidental
Signal	Tide Water
Sinclair	
St. Indiana	
Sun	
Union	
9 (64% of Total Group)	5 (36% of Total Group)

about equally its acquisitions of concessions between both hemispheres from the beginning and throughout the period. Their exploration policies were not altered following the imposition of import quotas in 1959.

Exhibit 2 shows evidence of a relationship between the company's crude position (defined as its self-sufficiency ratio) and the geographical pattern of exploration adopted. Having in mind to bring back any production they could develop abroad, the crude short firms went exploring first within close range in order to minimize the problem of logistics. But when the mandatory quotas enabled them to rely on a direct link between their close foreign production and their domestic refineries, they went where the biggest reserves were predicted to be, that is the Eastern hemisphere. Instead, the firms with no pressing need to feed their refineries, or without any refinery, aimed directly at the most promising geological areas whatever their geographical location. As early as 1952, for example, Continental stated that: "it is our management's belief that Continental is now well represented in almost every prospective area in the United States and Canada and that future acquisitions will generally be confined to smaller and more expensive spreads in highly promising areas."[1] A Continental executive recognized, however, some years later that: "because Continental planned to sell oil discovered abroad in the United States, some of our first efforts included exploration within the Western hemisphere, and hence, close to home."[2]

Intensity of the Exploration Effort Over Time. While the majority of the independents in this group—nine out of the fourteen—were well inside one

Exhibit 2
GEOGRAPHICAL PATTERNS OF EXPLORATION ABROAD STRATEGIC GROUP II – TOTALLY VERTICALLY INTEGRATED INDEPENDENTS 1945-1976

Company	Self-Sufficiency Ratio[1]	
	The Year Of Entry Abroad	In 1959
Western-then-Eastern Hemisphere Strategy		
Atlantic	.57	.71
Cities Service	.46	.45
Murphy	P	1.41
Phillips	1.04	.70
Signal	P	.96
Sinclair	.38	.37
Standard of Indiana	.52	.51
Sun	.47	.56
Union Oil of California	.75	.52
Worldwide Exploration Strategy		
Continental	1.35	.99
Getty	P	1.89
Marathon	2.63	.98
Tide Water	.53	.50

SOURCE: Companies' *Annual Reports.*

[1] A t test for two means (taking the average of each pair of figures) showed that the two foreign exploration strategies were associated with different self-sufficiency ratios (test significant below the 2.5% level).

standard deviation around the average intensity of 1.6 new country/concession entered/acquired each year, a broad range of intensity was displayed.

The diversity of exploration patterns over time allowed the independents to be categorized within a matrix 'level of exploration intensity–degree of continuity of exploration effort.' These results, summarized in Exhibit 3, were arrived at by applying the following procedure:

(1) firms with less than one new concession acquired each year in average during all the time they were active in foreign exploration were labelled 'low intensity' explorers;
(2) those with between one and two concessions acquired per year were considered 'average intensity' explorers; and
(3) those with more than two concessions per year 'high intensity' explorers,
(4) firms with exploration effort evenly spread over the period were

Exhibit 3
PATTERNS OF FOREIGN EXPLORATION OVER TIME TOTALLY INTEGRATED INDEPENDENTS (1945-1976)

	Level of Exploration Intensity (Total number of concessions acquired/number of years of active exploration)		
	Low Intensity (Below 1 conces./year)	Average Intensity (Between 1 and 2)	High Intensity (Above 2)
Cyclical (Effort Concentrated in Sub-Periods)	Sinclair Tide Water	Atlantic Sun	
Semi-Continuous (Uneven Effort Overall, but with Long Period(s) of Continuous Acquisitions)	Getty	Cities Service Phillips Signal St. Indiana Union Oil	
Continuous (Effort Evenly Spread Over the Period)		Murphy	Continental Marathon Occidental

SOURCE: Companies' *Annual Reports.*

called 'continuous' explorers;

(5) by opposition with 'cyclical' explorers, whose efforts were concentrated in separate periods of a few years each; and

(6) a category of 'semi-continuous' explorers was added. It covers either firms with long periods of active exploration separated by short periods of inactivity, like Union Oil of California and Getty, or firms with a first period of scattered activity followed by another period of continuous one, like Phillips, Signal and Standard of Indiana.

The variety of behaviors in the matter of exploration policy was witnessed during the interviews I made in the companies. In the early 1970s, Sun was not getting new concessions abroad because it was exploring on Sunray's properties, merged in 1968, and also because of the cash drain due to the development of its discoveries in Iran and in the North Sea.[3] In the case of Cities Service, it was the lack of dedication at the first push abroad that explains why no concessions were acquired between 1962 and 1967.[4] For a long time, Continental's policy was to put a few million dollars in each potential foreign country in order to get experience, and then see.[5]

I found no simple relationships between the intensity of exploration over time and either the company's size or its crude position.[6] However, when I divided the group in two between (1) the companies with an average self-sufficiency ratio (computed in the year of entry into foreign exploration and in 1959; see data in Exhibit 2) below .85,[7] and (2) the companies with an average ratio greater or equal to .85, a difference emerged. The crude short companies tended to be significantly less intensive–40%–in their exploration effort over time than the crude sufficient ones.[8] In addition, the two sub-groups were associated with significantly different sizes: in average, the crude short, low intensity explorers were twice as big as the crude sufficient, high intensity ones.[9]

It seems that, as time passed, low size was no more a barrier to entry as it was in the late 1940s, and that a satisfactory crude position was no more a deterrent from exploring abroad. In fact, though well balanced, the small independents were eager to grow and intensive foreign exploration was a means to achieve this objective, whereas the bigger independents, though crude short, put less emphasis on the exploration stage.

Integration Into Foreign Refining. The imposition of mandatory import restrictions by the U.S. government in 1959 played a key role in the formulation of the new strategies. It brought a change in the oil industry environment which left no alternative but to integrate downstream abroad or to remain an explorer-producer with a scope limited by the lack of a captive domestic outlet for any potential foreign production. Without import restrictions, Continental's whole international strategy would have been different.[10]

The evidence supporting this conclusion rests in the numbers of integration into foreign refining before versus after 1959. While thirteen independents in this strategic group had some production–though often small– outside the United States by 1959, only five of them integrated into refining abroad before that year. Yet Sinclair had to put up a refinery in order to get a concession in Venezuela.[11] It was also the case of Phillips who, "in compliance with its concession contracts with Venezuela, agreed in 1950 to construct a refinery . . . ."[12] The agreement between Saudi Arabia and Getty dated February 20, 1949, stipulated that: "when crude obtained by the Company reaches an average of 75,000 b/d, the Company is obliged to construct, within the neutral territory, a modern refinery with a minimum output of 12,000 b/d."[13] Construction was decided in 1957 when production was at 38,200 b/d.[14]

But thirteen of thc independents in this group integrated deliberately after 1959, and all but one–Occidental, a latecomer on the international petroleum scene–within five years. These observations are summarized in Exhibit 4.

Exhibit 4
IMPACT OF THE 1959 IMPORT CONTROLS ON DECISIONS TO REFINE ABROAD

Independents Going Into Refining Abroad:	
Before 1959	After 1959
Sinclair (1947)[1,2]	Getty (1959)
Phillips (1950)	Murphy (1959)
Sun (1951)	Tide Water (1959)
Cities Service (1956)	Continental (1960)
Getty (1957)	Marathon (1961)
	Signal (1961)
5	St. Indiana (1961)
	Sun (1962)
	Atlantic (1963)
	Sinclair (1963)
	Phillips (1964)
	Union (1965)
	Occidental (1968)
	13

[1] The dates refer to the year when the decision to go into refining was taken.
[2] The names underlined indicate that the refineries were based on production developed before 1959.

Among the thirteen independents who integrated into foreign refining after 1959, seven did so on productions they had developed before 1959 and that they until then shipped back into the United States. Among the remaining six, five waited until they got more substantial productions abroad, which were all below 10,000 b/d by 1959.

This behavior illustrates the point that, if the 1959 import controls led to a change of international strategy, it was by adding a stage (refining) to two previous ones (exploration and production), not by altering the timing of foreign production's development. A t test showed there was no difference in the quickness of the decision to go into refining whether they had previous production abroad or not.[15]

Procedures of Entry into Refining Abroad. Three procedures were used to enter into the refining stage after 1959: (1) processing arrangements with local refiners, (2) acquisition of existing refineries, and (3) construction from scratch. As shown in Exhibit 5, after adjusting for Continental's and Signal's quick changes of entry procedure and for Sun's limited integration, the three modes appear to have been used equally. This was not the case before 1959, where construction is the only observed pattern because the refining projects were located in countries lacking such facilities.

Although the test is not strong,[16] it seems that crude long or self-

Exhibit 5
PROCEDURES OF FIRST ENTRY INTO REFINING ABROAD AFTER 1959

Processing Arrangements	Acquisition	Construction
Atlantic	Getty	Marathon
Continental[1]	Occidental	Phillips
Murphy	St. Indiana	Tide Water
Signal[2]		Union
Sinclair		Sun[3]
5	3	5

[1] Continental started with construction in joint venture in Panama in 1960 but sold its participation in 1962 when construction was completed.
[2] Signal acquired two refineries the following year.
[3] Represents expansion of existing refinery.

sufficient companies chose to enter quickly into foreign refining through processing arrangement or acquisition, whereas crude short firms opted for the slower way of refinery construction. Acquisition procedure, which combines rapidity of implementation of the integration decision and stability of ownership, was not especially favored by either crude long or crude short firms.

Additional factors also played a role in these choices. In the case of Continental, for example, "the existence of excess refinery capacity in several countries of Western Europe offered a basis for obtaining processing deals on favorable terms."[17] In addition, "in its approach to European operations, Continental's management decided to postpone the building of a costly modern refinery until a degree of market security had been assured."[18]

In 1961, Marathon "reached agreement with Spanish and German interests to participate in the construction of two refineries . . . scheduled in about two years."[19] It is no coincidence that in 1961 Marathon's foreign production was only 6,807 b/d, but with Libyan fields going on stream, it jumped to 120,724 b/d in 1964. Occidental had concluded a process arrangement with Signal, with an option on its Belgian refinery which it exercised the day the first shipment of Libyan oil was made.[20]

Pecking Order of Integration into Refining and Marketing Abroad. The stages-of-development model presented in Chapter 4 hypothesized that integration into foreign refining would precede integration in foreign marketing. Two tests supported this hypothesis. A linear correlation test suggested a positive relationship between the timings of entry into refining and into retail marketing.[21] A rank correlation test based on the ranks of entry over the whole period both in refining and in marketing even showed that the first companies to go into foreign refining seemed also to be the first to go into retail marketing abroad.[22] Yet, as shown in Exhibits 6 and 7, a few independents, such as Atlantic, managed to get into marketing before refining.[23]

Exhibit 6
RELATIVE ORDER OF ENTRY INTO REFINING AND MARKETING ABROAD

#of Independents Going Into:[1]	During Whole Period	Of Which After 1959
Refining, then Marketing	6	5
Refining & Marketing at the same time	5	5
Marketing, then Refining	3	2

[1] Based on year in which the decision was taken.

A test showed that those independents going first into refining after 1959 were substantially more crude sufficient world-wide (as measured in 1959) than those going either first into marketing or into marketing and refining at the same time.[24] Because they were in or near crude surplus, the "refiners-then-marketers" were substantially quicker to decide going into refining abroad than the "marketers-then-refiners," which had not enough crude to feed their domestic refineries. The average lag time of investment into foreign refining between the two groups was almost two-and-a-half years.[25] This lag occurred because those that entered marketing first slowed down even more their actual entry into refining by choosing the construction procedure, which required a two- to three-year delay between the decision and the completion. By contrast, those which decided to get into refining and marketing at the same time favoured the acquisition procedure, for it combines rapidity and stability of implementation.

Investment and Divestment into Marketing Abroad. Although the independents had the choice for their first entry into retail marketing abroad between acquisition of existing service station chains and construction of new service stations, they went through acquisition in very large majority: twelve out of fourteen cases. Continental, for example, had the policy of buying the largest existing "true independent marketer" they could find in order to have from the beginning a sufficient sales volume to justify overheads as opposed to the scratch route of construction of service stations one by one.[26] In their following expansion, however, five of the twelve companies, who started with acquisition, used the construction procedure in addition.

Foreign marketing, though easy to enter, happened to be also the less stable of the independents' international strategy's components in terms of number of firms sticking to their original policies. As many as ten out of the

Exhibit 7
Multinationalization Through Verticalization: Evolution of International Strategies Over Time–The Integrated Independents (1945-1976)

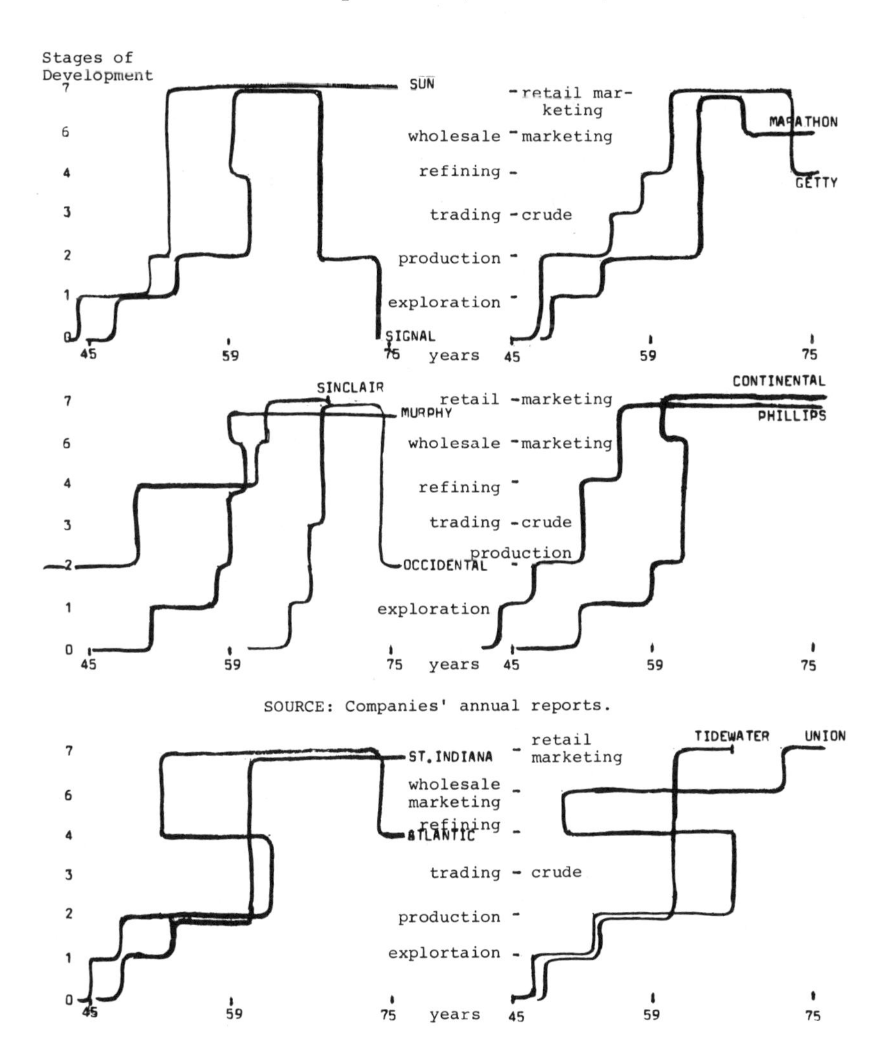

SOURCE: Companies' annual reports.

fourteen in the group divested more-or-less from direct marketing abroad within the last ten years of the period under study. These companies are listed in Exhibit 8. But among the remaining four, three reduced gradually the number of their foreign marketing outlets, especially since 1973, and the fourth one–Sinclair–was merged into Atlantic-Richfield.

Exhibit 8
DIVESTMENTS FROM MARKETING ABROAD

1962 – TIDEWATER sold marketing facilities in Europe.
1967 – PHILLIPS sold 95% interest in its Australian subsidiary.
1968 – SIGNAL sold its refining and marketing facilities in Europe to Occidental.
1968 – MARATHON sold retail outlets in Italy and West Germany.
1969 – MURPHY switched from company's service stations to dealers in Canada.
1970 – MURPHY sold its Swedish marketing subsidiary.
1973 – GETTY sold all marketing units in Europe to Burmah Oil.
1974 – ATLANTIC sold 450 U.K. service stations to CFP.
1974 – OCCIDENTAL sold most of its European outlets to Elf.
1975 – STANDARD INDIANA sold its petroleum assets in West Germany to a German firm.
1976 – UNION sold its Canadian refining and marketing properties to Husky Oil.

SOURCE: Companies' *Annual Reports.*

Union divested from foreign marketing because it was unprofitable.[27] For Occidental, profits were higher on brokerage than on retail marketing.[28] Standard of Indiana's policy was that marketing must stand on its performance.[29]

Reactions to the 1973 Oil Crisis. Measured by the number of foreign concessions acquired between 1974-76 as compared with the period 1971-73, the 1973 oil crisis clearly has had an influence on the integrated independents' foreign exploration policy. But it was less dramatic than could be expected. No one stopped exploring completely, as data in Exhibit 9 show. Signal is an exception, but it was not due to the oil crisis. The company's management had taken the decision to divest entirely from the petroleum business, domestically and internationally, just before the crisis.[30]

Among the six independents who reduced the pace of new concessions acquisitions, three remained, nevertheless, above their long-term trend (measured by the intensity of exploration between 1945 and 1976). Moreover, five others acquired more new concessions after the oil crisis than before. Continental's and Phillips' positions illustrate these contrasted behaviors. Continental is still exploring abroad even with an 85% split of profit in favour of producing countries' governments because it is still profitable to find oil abroad.[31] According to Phillips, on the contrary, an 85% government take is too high to allow a profit.[32] Murphy has chosen to explore now outside the O.P.E.C. countries, with an emphasis on English-speaking nations.[33]

Exhibit 9
IMPACT OF THE 1973 OIL CRISIS ON EXPLORATION POLICIES OF TOTALLY VERTICALLY INTEGRATED INDEPENDENTS

Company	Period 1974-1976*				Total # of New Countries Entered in 1971-73
	Industrialized Countries	Less Developed Countries		Total # of New Countries Entered	
		Total LDCs Including OPEC	Non-OPEC Countries Only		
Atlantic	1	2	2	3	5
Cities Service	6	8	4	14	9
Continental	9	7	4	16	24
Getty	2	3	3	5	11
Marathon	2	13	9	15	4
Murphy	1	3	3	4	6
Occidental	0	2	1	2	5
Phillips	1	6	5	7	9
St. Indiana	0	4	3	4	3
Sun	1	12	10	13	13
Union	4	7	6	11	4

SOURCE: Companies' *Annual Reports.*
*Concessions acquired in stated period and stated categories.

Degree of Balance in the Foreign Vertical Integration Chain Over Time. I compared the degree of balance reached when the vertical integration process was started abroad with the one achieved at the same time in the United States. I assumed that, whatever their domestic balance, the independents aimed at a positive balance abroad.

Three measures of the degree of vertical integration were used in the analysis:

(1) the ratio between own production and refinery run, known as "self-sufficiency ratio" ('production/refining' hereafter);
(2) the ratio between refinery runs and refined products sold ('refining/marketing' hereafter); and
(3) the ratio between own production and refined products sold ('production/marketing' hereafter).

These are all self-sufficiency ratios, but they emphasize different points of the vertical integration chain. The ratios were computed from data of production, refining, and marketing published by the firms in their annual reports. Because of the lack of standard format in the publication of these statistics, the ratios could be computed for only nine of the fourteen totally integrated independents.

As summarized in Exhibit 10-B, in twenty-five cases out of twenty-seven possible (nine companies x three self-sufficiency ratios), the observed starting balance positions were positive as expected. The rationale was to be profitable from the beginning. As Continental put it: "The performance of the established international oil companies spoke in favor of vertical integration. It appeared to us that those companies with the better balanced operations in the Eastern Hemisphere, as indicated by the relationship of their own crude production to refinery runs, had the better performance over a sustained period of time as evidenced by their earnings per barrel of crude produced."[34]

Using the same three self-sufficiency indicators, I then compared the evolution over time of the vertical integration's balances domestically and internationally. The test of the convergence or divergence of the vertical integration process through time was the correlation coefficient between domestic and foreign time-series of self-sufficiency ratios. A positive correlation coefficient was associated with a 'converging' pattern of integration, whereas a negative one was associated with a 'diverging' pattern. The results are presented in Exhibit 11. There does not seem to exist any clear relationship, since eleven converging patterns of balance were observed compared with nine diverging patterns and seven cases of absence of influence. However, when only the correlation coefficients significant at 5% are taken into consideration (see Exhibit 11, columns 'a' and 'c'), a clearer picture emerges. All the independents focusing on just one indicator of integration balance displayed a diverging pattern; that is, for example, if a firm had a decreasing crude position in the United States, it managed to have an increasing one abroad. On the contrary, all the independents who emphasized at least two indicators of integration balance showed converging patterns of balance; that is, if a firm monitored both 'its production/refining' and 'refining/marketing' self-sufficiency ratios, these tended to follow a parallel evolution both at home and abroad.

I. A Case Study: Atlantic/Atlantic-Richfield

Atlantic has been active abroad without interruption for sixty years. Although not a leading company for a long time, the point is that it remained remarkably steady on the principle of its presence abroad even through World War II. As pointed out in Chapter 5, whereas twelve independents had some kind of experience abroad in the 1920s and 1930s, only six of them remained active by 1939-40. But just two—Sun and Atlantic—kept expanding throughout the war.

Exhibit 10
RELATIONSHIPS BETWEEN DOMESTIC AND FOREIGN SELF-SUFFICIENCY RATIOS

A. Detailed Results

Domestic Vertical Integration Chain	Foreign Vertical Integration Chain: Production/Refining (Own Production/ Refinery Runs)		Refining/Marketing (Refinery Runs/ Products Sold)		Production/Marketing (Own Production/ Products Sold)	
	High (ratio $>$ 1.)	Low (below 1.)	High (above 1.)	Low (below 1.)	High (above 1.)	Low (below 1.)
High: ratio $>$ 1.	Getty Marathon		Getty Marathon Sinclair St. Indiana Sun	Atlantic Union	Getty Marathon	
Low: ratio $<$ 1.	Atlantic Continental Phillips Sinclair St. Indiana Sun Union		Continental Phillips		Atlantic Continental Phillips Sinclair St. Indiana Sun Union	

B. Summarized Results

	H	L
H	2	
L	7	

	H	L
H	5	2
L	2	

	H	L
H	2	
L	7	

☐ means expected patterns. SOURCE: Data derived from companies' *Annual Reports.*

Exhibit 11
RELATIONSHIPS BETWEEN EVOLUTION OF DOMESTIC AND FOREIGN BALANCES

	Converging Integration's Balances		Diverging Integration's Balances		No Relationship
	Correlation Coefficient Between Domestic and Foreign Ratios Significant:				
Self Suffi-ciency Ratios	at 5% [a]	at 20% [b]	at 5% [c]	at 20% [d]	not at 40% [e]
Own Production/ Refinery Runs	Marathon Phillips	Continental Getty	St. Indiana	Atlantic	Sinclair Sun Union
Refinery Runs/ Products Sold	Continental Sun		Atlantic Phillips Sinclair	St. Indiana Union	Getty Marathon
Own Production/ Products Sold	Continental Marathon Phillips Sun	St. Indiana		Atlantic Getty	Sinclair Union

SOURCE: Analysis based on data published in companies' *Annual Reports.*

Atlantic's policy of increasing its holding of foreign acreages culminated in 1941, but in the following years the emphasis shifted to increasing the U.S. holdings: three-fold between 1941 and 1945. However, "not unmindful of the trends,"[35] in 1943 Atlantic continued geological and geophysical exploration in Venezuela, Cuba, and Haiti. In February 1944 it concluded a drilling agreement with Pantepec in Venezuela. Atlantic was to drill eight exploratory wells before the end of March 1947; by then, it had thirty days to buy a 50% interest for $4 million. When the hostilities ended, Atlantic did not waste any time in resuming its aggressive policy of leaseholds acquisition abroad. On May 31, 1945 it purchased a 50% interest in a Pantepec block located in Venezuela for $2 million. The purpose was to maintain a policy formulated in 1943 of augmenting the company's crude reserves.[36] To make the best use of the properties acquired, prompt, extensive geological and geophysical exploration and costly testing through drilling were undertaken. It paid off, for in the last quarter of 1946 production was found, and in September 1948 Atlantic imported into the United States its first cargo of Venezuelan crude. In the meantime it had exercised its entire option to purchase 50% of the corresponding concession held by Pantepec.

This offensive on foreign exploration implied an important financial effort. In 1947, total expenditures for this purpose were three times as large as for any year before 1944, and nearly five times similar annual expenditures before 1940. Not only half of this total was spent abroad versus only half domestically, but it raised also a resource allocation problem. In 1947, Atlantic's management acknowledged that the volume of crude it produced in the United States was only slightly higher than in 1946 primarily because of the need to devote a large share of available funds to foreign exploration.[37]

While some exploration activities were under way in Cuba, Haiti, Nicaragua (since 1946), and Guatemala (since 1948), they did not last long. Concessions in Haiti were relinquished in 1947; efforts were discontinued in Guatemala and Nicaragua and suspended in Cuba in 1949. Such a high turnover occurred because the main effort was made in Venezuela. When it discovered oil on a concession owned by Pantepec, Atlantic exercised its option on half. Since it was not possible to get new concessions from the Venezuelan Government, in 1947 Atlantic moved to acquire a half interest in one lease owned by Barnsdall South American Company. The following year, it took a two-third participation in the Guasere Oil Company, which held concessions in western Venezuela.

Success in Venezuela, however, raised what the management seemed to construe as a 'moral' problem at the same time it brought needed crude in the U.S. Was Atlantic contributing to the threat posed to part of the U.S. oil industry by the fast-rising imports? In 1949, over 4 million barrels from the Venezuelan production were delivered to Atlantic's refinery in Philadelphia.

However, it was the 1,200 b/d–ten times less–imported by Atlantic from Kuwait the same year that were noticed by Congress in a report on the "Effects of Foreign Oil Imports on Independent Domestic Producers."[38] As, on one hand, one of the first independents to get any significant production abroad and, on the other hand, one of the most crude short ones even with this addition, Atlantic's position on the issue was one of political commitment.

"Within the industry, the question of whether imports supplement or supplant domestic production is a matter of semantics and not of economics. In my opinion,[39] there is now little doubt that the potential oil resources of the rest of the world are greater than those of the United States. What I am afraid of is that, unless America encourages its oil industry to enter the foreign field, this abundant supply might become firmly controlled by foreign operators, even by potential enemies."

But, by 1950, for Atlantic it had become a question of survival more than anything else. To fight low-price foreign crude competition, domestic production had been restricted by state regulatory authorities to about 82% of the efficient capacity. Therefore the availability of Venezuelan crude to supplement Atlantic's domestic production was of great importance, particularly in view of a more balanced integration which had been a management concern for the past years.[40]

Still, that was not enough to solve the supply problem. As a result Atlantic got more involved in the import question. In December 1950, Atlantic announced that it had entered into a contract with Gulf to buy approximately 12,000 b/d of Kuwait oil for a period of five years beginning in 1951.[41] By then two of the three supertankers ordered by the company to be used to transport crude from the Persian Gulf to the Philadelphia refinery were delivered.

After the end of World War II, not only did Atlantic implement an aggressive exploration policy abroad aimed at quick production, but at the same time it resumed an international marketing activity that had been an important element of its pre-war strategy. Quickly the Foreign Marketing Department was able to make substantial steps toward a return of pre-war conditions in most of the war-ravaged countries served by Atlantic. By the end of 1946, it marketed again in Brazil, Belgium, South Africa, Portugal, West Africa, and to a lesser extent in North Africa and the Netherlands; that is, practically, the same list of countries as ten years before.

Atlantic's international marketing strategy was, however, to be challenged by changing environmental conditions. First, in 1948, the growing demand in Europe had to be met without affecting the company's ability to serve the also growing demand in the U.S. Then in 1952, it had to adapt to a basic change in the method of retailing gasoline in South Africa. Service stations selling many brands were replaced by service stations handling only one. In spite of the strong competition, Atlantic succeeded in securing some of those new outlets.

In 1954, however, Atlantic sold all of its Eastern hemisphere marketing subsidiaries to British Petroleum. "We have withdrawn from a marketing area where we were at a disadvantage against competitors having more available crude supplies and refining capacities."[42] Atlantic thus learned that it was desirable for a petroleum company willing to get involved in marketing abroad to be integrated up-stream both in production and refining. The irony of the argument is that it sold at the same time a 5% interest in Esso–France–which had refining facilities and was within a protected market that no other U.S. independents could ever penetrate later on–six months before the Parentis discovery.

The operation was, however, not a complete opportunity loss, although it cost Atlantic immediately 11% of its dollar sales. The major part of the properties was to be paid over a period of years by deliveries by British Petroleum of 12,000 b/d from Kuwait, therefore taking the relay of the Gulf contract of 1950. The part in cash was devoted to the realization of the long-time objective of making the company more self-sufficient. This was done through two successive moves.

Early in 1955, the U.S. independents were offered the opportunity to share 5% of the Iranian Consortium.[43] Inside directors of Atlantic's board were against the project, which required more the expertise of a broker than of an oil company, but H. Supplee, President since 1952, argued convincingly that it was the cheapest way to get additional foreign production. Atlantic bid for 2%, but eventually got 1/12 of the 5%. Quite curiously, however, no mention of it was ever made in the company's *Annual Report* until ten years later. Still, the investment paid out in eighteen months,[44] but that too went unnoticed.

The next move was much more publicized, although both were key in the development of the independents as a group of international oil companies. In 1956, Venezuela opened for bid concessions in Lake Maracaibo. Atlantic got a 45% interest in two blocks covering 27,000 acres. But it paid a bonus of $10 million because the concessions were located in a region of prolific producing areas. In the 1957 bidding round, Atlantic spent $28 million in bonus.

The 'reentry' into Venezuela marked the beginning of an intensive exploration policy. In the late 1940s, foreign exploration was concentrated mainly in Venezuela. During the first half of the 1950s no additional concessions were acquired anywhere. But between 1955 and 1959 a dozen countries were entered or reentered. Still the bulk of the effort was directed toward the Western hemisphere. The two attempts in the Eastern hemisphere were unsuccessful, although for different reasons. For two years Atlantic tried to get into Turkey. Eventually it got a one-third interest in 3 million acres in 1957. But one year later, exploration was abandoned: a substantial quantity of gas had been found, but no oil. In Syria, oil was discovered for the first time in 1956 by J.W. Marshall, an American independent operator who became the same year associated with Atlantic. But the Syrian government canceled the exploration permit in 1958 on grounds that Marshall had not carried out drilling obligations and had transferred an interest to Atlantic without a prior government approval. A total of $4 million had been invested in the venture, of which Atlantic had put up about $3.5 million.[45] But significantly enough, the company did not try to get into Libya at the 1955 bidding round because it was busy enough with its share of production from the Iranian Consortium.[46]

Atlantic was more successful in the Western hemisphere. Sizeable discoveries were made in Lake Maracaibo in 1957 and 1959, and in Alberta in 1957. Truly Atlantic was exploring in areas where oil had been already discovered; but this was in accordance with the company's policy to increase its self-sufficiency in raw materials. In 1959, Atlantic's management claimed that it had attained a potential equilibrium in that regard: the self-sufficiency ratio rose from 52% in 1955 to 71% in 1959.[47]

At first, Atlantic was sympathetic with the U.S. program of voluntary import restriction: "We believe the problem to be relatively temporary."[48] By that it was meant that demand outpacing domestic production made inevitable increased imports. When in 1957 Atlantic's quota was reduced by nearly 10,000 b/d, the management did not react too harshly because its Venezuelan production still represented less than a quarter of that quota. But, starting in 1959, Atlantic was obliged to buy additional import quotas in order to utilize its growing Venezuelan production. In 1962 tighter restrictions and worldwide crude surplus worsened the situation and led to some changes of strategy affecting the Venezuelan operations.

In 1964 some scattered producing properties worth 19,600 b/d were sold. A long-term refining agreement was signed for the processing of 20,000 b/d of Atlantic's Lake Maracaibo crude at a refinery in Venezuela operated by Mobil. Arrangements also were made for the purchase by the refiner of substantial quantities of additional crude oil from Atlantic. The following year, Atlantic received its first deliveries of products under this refining agreement and sold them in various world markets.

In terms of the model presented in Chapter 4, within a two-year span, Atlantic shifted from Stage 3A (importing foreign production into the U.S.) to Stage 3 (selling it on the international market), then integrated Stage 4 (refining its foreign production abroad) and Stage 5 (selling the products at the refinery's doors). The reason why Atlantic reacted with a substantial time lag to the import restrictions is because its Venezuelan production did not exceed its quota until 1962. In other words, it is not until Atlantic was compelled to find new outlets for its foreign production that it took steps to integrate downstream. Yet it did so in the most flexible way through a long-term refining arrangement instead of heavy capital outlays put in the construction of a new refinery.

The early 1960s witnessed also a change in the geographic distribution of the foreign exploration effort. Whereas between 1955 and 1959 the bulk of new concessions acquisitions was made in the Western hemisphere, the ratio was completely reversed in favor of the latter between 1960 and 1965. Then again, Atlantic was targeting the big fields by going in such promising areas from a geologist's standpoint or even proved areas such as Libya (1960), Sahara (1960), Australia (1961), Iran (1964), and the North Sea (1964, 1965).

From 1966 to 1970, Atlantic did not acquire any new concessions abroad, for it had inherited the foreign concessions portfolios of Richfield and Sinclair by merging with them respectively in 1966 and 1969. Richfield added holdings in Iran, Morocco, the North Sea, Australia, Guinea, Canada, and Colombia; Sinclair brought leases in Argentina, Libya, Saudi Arabia, and Indonesia. The first task of the new group was to consolidate existing concessions rather than to think of expanding. Additions to the exploration portfolio were not the only impact of the two successive mergers. As shown in Exhibit 12, the effect was to increase the size of Atlantic abroad. In particular, the merger between Atlantic-Richfield and Sinclair enhanced further its vertical integration commitment abroad. This effect is especially true as far as refining and marketing are concerned. Sinclair brought refineries, though small, in Venezuela and Peru and larger processing arrangements in Europe, as well as service stations in Puerto Rico, the Bahamas, El Salvador, the Dominican Republic, and mostly in Belgium and the United Kingdom. By comparison Richfield, not integrated abroad, brought only an interest in the Iranian Consortium's refinery of Abadan and an export volume from the United States, which was on a decreasing trend.

Reversing a ten-year-old policy in 1964, Atlantic reentered foreign marketing by building and then acquiring service stations in the United Kingdom. In 1966 it claimed 200 retail outlets. In 1969 it got 100 from Sinclair and in 1972 it acquired 435 more from Burmah Oil. But in September 1974 all the marketing facilities in the U.K. were sold to the British subsidiary of Compagnie Française des Pétroles (C.F.P.). No explanation was given by the seller, but the buyer's reasons provide a clue. C.F.P. claimed that "oil reserves discovered in the North Sea are bound to help stabilize the British market."[49] Atlantic had no oil production in the North Sea and therefore was at a competitive disadvantage. The closest production it had–20,000 b/d out of Libya–had been nationalized in February, 1974.

The company's principal source of foreign crude was Iran, with a share in the Iranian Consortium and a discovery made offshore in 1965. But

Exhibit 12
IMPACT ON ATLANTIC OF MERGER WITH RICHFIELD AND SINCLAIR

	Atlantic's Rank Among the U.S. Independents in 1965 Based on Quantities Produced, Refined or Sold	Atlantic + Richfield/ Atlantic	Atlantic-Richfield's Rank in 1966 After Merger	Atlantic-Richfield- + Sinclair/ Atlantic-Richfield	Atlantic Richfield's Rank After Merger with Sinclair in 1969
Foreign Production	9	+31%	4	+102%	3
Foreign Refining	9	+18%	8	+157%	6
Foreign Marketing	9	+141%	5	+153%	5

SOURCE: Data from Atlantic's, Richfield's, Sinclair's, and Atlantic-Richfield's *Annual Reports.*

Exhibit 13
SUMMARY OF THE MAIN FINDINGS*

Causes	Patterns			
	Exploration			
	Intensity of Exploration	Geographical Location	Refining	Marketing
1959 Import Controls	increased	shift to Eastern Hemisphere	triggered integration	
Company's Size	bigger, lower intensity	no influence	the biggest, the first to integrate	no influence
Company's Self-Suffi-ciency	crude short: low intensity; crude sufficient: high intensity	crude short: Western-then-Eastern Hemisphere; crude long: Worldwide strategy	crude long: quicker to integrate	crude short: integrated first into marketing
1973 Oil Crisis	some companies increased; some decreased	avoid OPEC, but not other LDCs		

* Totally vertically integrated independents.

the import quotas system had been eliminated in May 1973, therefore providing a new incentive for a domestic crude short company such as Atlantic to bring back into the U.S. its foreign production. The 1973 oil embargo disturbed this company policy to the point of forcing the formulation of a new international strategy. Atlantic was to become domestically self-sufficient, no longer world-wide sufficient.[50] Foreign crude was to supply foreign markets where it was profitable.[51] Accordingly because returns on investments were not satisfactory, not only did it drop from marketing in the U.K., but refining arrangements in Europe were reduced to one-sixth between 1974 and 1976. For the same reason and also because the Canadian government set export restrictions toward the United States, Atlantic-Richfield divested entirely from Canada in 1976.

Conclusion

As for the other strategic group, the main findings concerning the totally vertically integrated independents are summarized in a synthesized way, with an emphasis on what caused the observed patterns.

NOTES

[1] Continental, *Annual Report 1952.*

[2] Harvard Business School, "Continental Oil Company," Case study, 1967, p. 6.

[3] Information obtained during an interview in Sun.

[4] Information obtained during interviews in Cities Service.

[5] Information obtained during interviews in Continental.

[6] The Pearson correlation coefficient was .05 between intensity of exploration and companies' size in 1959; it was .11 (significant at only 35%) between intensity and companies' self-sufficiency ratios in 1959.

[7] During interviews, several companies considered a ratio of .85 between own production and refinery runs as close to self-sufficiency.

[8] A t test for two means showed the difference to be significant below the 10% level.

[9] A t test for two means showed the difference to be significant at the 5% level.

[10] Information obtained during interviews in Continental.

[11] Information obtained during interviews in Atlantic-Richfield.

[12] Phillips, *Annual Report 1950.*

[13] Pacific Western Oil Company (renamed Getty), *Prospectus,* June 1949.

[14] Getty, *Annual Report 1957.*

[15] The t test was significant only at the 50% level, which means that there is as much chance for the average delays of reaction to be different or the same among the two subgroups.

[16] The t test for two means was significant at the 15% level.

[17] Harvard Business School, "Continental Oil Company," op. cit., p. 11.

[18] Ibid, p. 12.

[19] Marathon, *Annual Report 1961.*

[20] This information was obtained during an interview in Occidental.

[21] The linear correlation coefficient was .51 (significant at almost 2.5%), taking the year of completion when refineries were constructed; r = .44 (significant at almost 5%), when the year of decision was considered instead.

[22] The Spearman's rank correlation coefficient was .488 (significant below the 5% level) when based on effective entries, i.e., year of completion when construction was used. If based on year of decision, r = .422 (significant slightly over 5%).

[23] See Section 3 in this chapter for a detailed survey of Atlantic's strategy abroad.

[24] The t test for two means was significant below the 2.5% statistical level.

[25] A t test showed this lag as being significant below the 10% level.

[26] This information was obtained during interviews in Continental.

[27] This information was obtained during interviews in Union.

[28] This information was obtained during an interview in Occidental.

[29] This information was obtained during interviews in Standard of Indiana.

[30] This information was obtained during an interview in Signal.

[31] This information was obtained during interviews in Continental.

[32] This information was obtained during interviews in Phillips.

[33] This information was obtained during interviews in Murphy.

[34] Harvard Business School, "Continental Oil Company," op. cit., p. 11.

[35] Atlantic, *Annual Report 1943.*

[36] See Atlantic, *Annual Report 1943* and *Annual Report 1945.*

[37] Atlantic, *Annual Report 1947* and *Annual Report 1948.*

[38] U.S. Congress, op. cit., 1949.

[39] Atlantic's President, R. H. Colley, March 22, 1950, quoted in *Annual Report 1949.*

[40] Atlantic, *Annual Report 1947* and *Annual Report 1950.*

[41] U.S. Senate, "The International Petroleum Cartel," 1952, p. 145.

[42] Atlantic, *Annual Report 1954.*

[43] On the Iranian Consortium, see R. B. Stobaugh, "Evolution of the Iranian Oil Policy: 1925-75," op. cit.

[44] This information was provided during interviews made in the company.

[45] Barrow, G., *International Petroleum Industry,* op. cit., p. 249.

[46] This information was obtained during interviews in Atlantic-Richfield.

[47] See Atlantic, *Annual Report 1959.*

[48] Atlantic, *Annual Report 1956.*

[49] Compagnie Française des Pétroles, *Annual Report 1974.*

[50] Information provided during interviews by author in the company.

[51] Ibid.

CHAPTER VIII

CONCLUSION

I. Summary of the Main Findings

This study has established the existence and strength of relationships between the U.S. independents' size and crude position and the main development steps of their international strategies.

A firm's size—measured as dollar sales relative to Exxon's—was a factor in the decision to go abroad. The biggest among the independents displayed a definite propensity to be the first abroad in the years immediately following the end of World War II. Because of the lumpiness of the first investment into foreign exploration, big companies could afford the additional burden whereas it was a barrier for the small companies.

Size was also a decisive factor when the question of whether to integrate downstream abroad or not came up. The firms choosing to integrate vertically tended to be bigger than those choosing not to. And if there were a few exceptions to this rule, they were not both ways: none of the big independents stopped at production. Besides, the few small independents that integrated abroad were crude long, whereas all those that stopped at foreign production were small and generally crude short. The lumpiness of the first investment into refining in a foreign country was such that only the big firms could bear the cash drain it represented. It was only if they had substantial or fast-growing production overseas that the small independents could skip the barrier of entry into foreign refining.

The firm's crude position—measured by its self-sufficiency ratio—has been a factor also in the main decisions pertaining to the independent's international strategy. This study was able to illustrate the point in several situations.

The first entries into foreign exploration after the end of World War II were dictated by crude positions. Between 1946 and 1951, in particular, the independents going abroad were generally crude short whereas those still domestic in scope were generally crude long. Crude short firms had, by definition, a more compelling need to feed their refineries than the crude sufficient ones.

Crude positions were also found in close relationship with the geographical range of the foreign exploration policies. In both strategic groups, the crude short firms, because they were in pressing need to find crude oils for their refineries, began by exploring in the Western Hemisphere. On the contrary, the crude long firms had a worldwide scope of exploration from the start because they were more concerned with geological opportunities than with logistic problems.

Crude positions of the independents, which integrated totally abroad, also commanded the procedure and timing of entry into refining as well as the pecking order of entry into refining versus marketing. Crude long firms were quicker to go into refining abroad because they had to find an outlet for their growing foreign production. The crude short firms tended to get into marketing first, waiting to have secured a stable market before undertaking the lump investment of a refinery abroad.

Changes in the environment of the international petroleum industry triggered two main moves by the independents. The beginning of import in the United States of low-cost crude oils from the Middle East by the American majors jeopardized the independents' markets in the U.S. and therefore compelled them to go and find their own sources of low-cost crude.

The imposition of import restrictions in 1959 triggered a wave of integration into refining and marketing. Because they could no more import back in the United States the totality of their foreign production, the biggest independents resolved to build up a totally vertical integration chain close to their foreign production.

The 1973 oil crisis does not seem to have triggered drastic reactions in terms of exploration policy. If the independents tend to avoid the O.P.E.C. countries, they are still committed to explore in less developed countries, for that is where the bulk of the yet undiscovered reserves of crude oil can be found.

Beyond the different patterns adopted on specific policy matters–although never more than two or three–what remains is the perenniality of the strategies over time. All the independents began with a sourcing strategy in the late 1940s. They went abroad partly because they were threatened by the incoming of cheap crude oil from the Middle East that they had no part in producing, partly because the cost of drilling domestically had gone up; in short, because there were more profits to be made abroad.

When the U.S. government imposed stiff restrictions on the oil companies' foreign sourcing strategy in 1959, some decided just to live with the constraints by sticking to their initial strategy. More decided to take advantage of the same constraints by going into manufacturing and marketing abroad. By adding manufacturing to sourcing, they in effect went multinational through vertical integration. Once these basic choices were made, companies in great majority adhered to them until the end of the period under study.

II. Implications for the Oil Industry

With very few exceptions, the perenniality of strategy emerged as a strong characteristic of the group studied. To extrapolate a thirty-year trend in the future is therefore a straightforward task. Although it is speculation, I will expect no drastic change of international strategy among the independents surveyed. The partly integrated companies will remain that way. I will expect the same stability in the other strategic group of the totally vertically integrated companies.

Some external disturbances may, however, come into play. If the U.S. government were to reimpose import quotas in order to reduce the dependence on foreign oil, major effects would follow. I would anticipate reactions from the companies similar to those that followed the 1959 decision. Compelled to separate domestic and foreign business, I would look for some of the explorers-producers to integrate downstream as well as for some integrated companies to become more integrated abroad.

The recurrence of the same variables as explanations for the main strategic decisions warrants their use as predictors of behavior in case of a new import quotas system. I would expect the biggest among the explorers-producers more likely than the smallest to become more multinational through further integration. Also I would expect the crude long firms or self-sufficient worldwide to be more likely candidates than the crude short ones. As for the already integrated independents, I would assume an additional emphasis put on a separate balance of the domestic and foreign operations. This objective would likely be implemented by paralleling abroad the domestic pattern of vertical integration balancing.

Since the 1973 oil crisis affected the independents to a lesser extent than could be expected, I would assume them to hold to the compromise that they seem to have adopted: that is, to be present in the less-developed countries, but to avoid the O.P.E.C. members.

This study also has implications for two other actors in the international oil industry: the U.S. government and the foreign governments. These implications are, in fact, the same as for the companies, but viewed from a different standpoint.

If, in order to achieve its goal of nationwide self-sufficiency, the U.S. government were to restrict imports, they should be aware of the likely impact on the companies' global strategies. If one effect is to send the oil companies once more back abroad, it will be in direct contradiction with the objective of domestic self-sufficiency, because if companies are forced to allocate additional resources for the foreign vertical integration chain, there is that much less for the domestic operations.

Foreign governments face two implications from this study. So far, few less-developed countries have attracted downstream investments from the independents. It is rather unlikely that they will do better in the future in view of the increased risk of nationalization. On the contrary, highly stable exploration objectives over time could be used to the advantage of the LDCs, provided some precautions are taken. In order to profit from the follow-the-leader behavior exhibited by the independents in the management of their exploration portfolios,[1] LDCs could try to attract consortiums of companies instead of just individual bidders. Doing so, they would give the companies enough power to counter-balance the nationalization threat by one government on an isolated company.

III. Contribution to Existing Bodies of Knowledge and Further Lines of Research

The ultimate academic purpose of this research is both a progress in the state of the art and the building up of a basis for further research. As far as the energy field is concerned, this study covers new ground through its focus on an overlooked group of firms, the international independents. The international strategies of a population of twenty-two American independents have been analyzed for a period of about sixty years.

The international business field should benefit from research focusing on overall international strategy, not merely the first decision to go abroad. The complexity of the successive investment decisions over a thirty-year period could be analyzed in terms of simple factors such as a company's size, crude position, past experience, and concern for profit. The business policy area should profit as well. The traditional model of strategic decisions taken, given some opportunities, after review of the company's strengths and weaknesses is incomplete, for it does not take into account the basic stability of a company's strategy over a long period of time.

The studies on the effect of environment on management should also incorporate the findings of this research. Firms are more likely to react to change in their working environment if they directly affect their domestic markets, less likely if they affect only the conditions overseas.

One objective of this research was to evaluate the possibility of high stability of strategy over a long period of time. It would be interesting to replicate such an inquiry for other industries. This should be easier in the case of raw material industries, for the model of multinationalization through verticalization tested in this study has been conclusively supported.

The observation of two strategic groups supports previous studies[2] that concluded with the existence of asymmetrical industries rather than symmetrical ones. The interest of this study is that it is the first attempt to test the validity of the concept on the international portion of an industry. Such a test could be repeated on other industries.

This study did not venture into financial analysis. One of the main findings suggests, however, an obvious line of research. Do different strategic groups mean different performances? McLean and Haigh were able to show the stabilization of profitability brought by domestic integration.[3] But the fact that some of the integrated independents divested more-or-less extensively from foreign marketing in the late sixties and early seventies for lack of profitability[4] suggests that verticalization may not be the economically efficient process claimed, at least abroad.

One additional line of research would be to test Chandler's findings in an international situation. Were the changes in the international strategies observed in this study followed by changes in the international structures? This study was not designed to go into the depth necessary to gather data on the companies' administrative systems. Partial evidence showed, however, that changes of structures paralleled the steps of verticalization abroad. So far one study has related the adoption of a given international structure to the degree of involvement abroad as measured by the ratio foreign sales/worldwide sales.[5] There is an opportunity now to use a better proxy: the stages of development of the firms' international strategies.

This study, being a first attempt at analyzing each independent's strategy, did not pay much attention to the interactions among them. Of the sixteen companies I visited, only one acknowledged that it was watching the competition. I presented, however, some evidence suggesting a follow-the-leader behavior in foreign exploration.[6] Previous research, although based on a different sample, showed similar patterns in foreign refining investments.[7] The frequency of merger talks and mergers concluded between the U.S. independents, particularly in the 1960s, strikes also as a clue of interaction. A research aimed at highlighting oligopoly behaviors would be worthwhile now that the individual international strategies have been analyzed.

NOTES

[1] See Appendix 5 for an analysis of interactions among independents in exploration abroad between 1945 and 1976.

[2] See Chapter IV, section III.

[3] McLean and Haigh, op. cit., 1954.

[4] Tentative evidence gathered from interviews in the companies.

[5] Stopford and Wells, op. cit.

[6] See Appendix 5.

[7] Knickerbocker, op. cit.

APPENDIX 1–SELECTED BIBLIOGRAPHY ON INTERNATIONAL PETROLEUM

Abir, Mordachai. *Oil, Power and Politics: Conflict in Arabia, The Red Sea and the Gulf.* London: F. Cass, 1974.

Adelman, Morris A. "The Multinational Corporation in World Petroleum," in Charles P. Kindleberger, *The International Corporation.* Cambridge, Mass.: M.I.T. Press, 1970.

——. *The World Petroleum Market.* Baltimore: John Hopkins University Press, 1972.

Barrows, Gordon H. *International Petroleum Industry.* New York: International Petroleum Institute, 2 vols.: 1965, 1967.

Brannon, Gerard M. *Energy Taxes and Subsidies.* Cambridge, Mass.: Ballinger, 1974.

Business History Review. *Oil's First Century.* Boston, Mass.: Graduate School of Business Administration, Harvard University, 1960.

De Chazeau, Melvin G. and Alfred E. Kahn. *Integration and Competition in the Petroleum Industry.* New Haven, Conn.: Yale University Press, 1959.

Choucri, Nazli. *International Politics of Energy Interdependence: The Case of Petroleum.* Lexington, Mass.: Lexington Books, 1976.

Continental Oil Company. *Conoco: The First One Hundred Years: Building on the Past for the Future.* New York: Dell Publishing Company, 1975.

Donaldson, Lufkin & Jenrette, Inc. *The International Oil Industry: The Improving Environment Abroad Increases Earning Potential for International Oils.* New York, 1966.

Duchesneau, Thomas D. *Competition in the U.S. Energy Industry.* Cambridge, Mass.: Ballinger, 1975.

Fanning, Leonard M. *Foreign Oil and the Free World.* New York: McGraw-Hill, 1954.

Fesharaki, Fereidun. *Development of the Iranian Oil Industry: International and Domestic Aspects.* New York: Praeger, 1976.

Frankel, Paul H. *Mattei, Oil and Power Politics.* New York: Praeger, 1966.

Gabriel, Georg. *The Gains to the Local Economy from the Foreign-Owned Primary Export Industry.* Unpublished Doctoral Dissertation, Graduate School of Business Administration, Harvard University. Boston, 1967.

Ghadar, Fariborz K. *The Evolution of O.P.E.C. Strategy.* Lexington, Mass.: Lexington Books, D.C. Health and Co., 1977.

Jacoby, Neil H. *Multinational Oil: A Study in Industrial Dynamics.* New York: MacMillan, 1974.

Krueger, Robert B. *The U.S. and International Oil: A Report for the Federal Energy Administration on U.S. Firms and Government Policy.* New York: Praeger, 1975.

Mancke, Richard B. *Squeaking By: U.S. Energy Policy Since The Embargo.* New York: Columbia University Press, 1976.

McDonald, S. "U.S. Depletion Policy–Some Changes and Likely Effects." *Energy Policy,* (vol. 4) March 1976, pp. 56-62.

McLean, John G. and Robert W. Haigh. *The Extent of Vertical Integration in the Oil Industry in 1950.* Boston: Division of Research, Graduate School of Business Administration, Harvard University, 1952.

——— and ———. *The Growth of Integrated Oil Companies.* Boston: Division of Research, Graduate School of Business Administration, Harvard University, 1954.

Medvin, Norman. *The American Oil Industry: A Failure of Anti-Trust Policy.* New York: Marine Engineers Beneficial Association, 1973.

Mikesell, Raymond F. and Hollis B. Chenery. *Arabian Oil, America's Stake in the Middle East.* Chapel Hill, North Carolina: University of North Carolina Press, 1949.

Mikesell, Raymond F., et al. *Foreign Investment in the Petroleum and Mineral Industries: Case Studies in Investor-Cost Relations.* Baltimore: Johns Hopkins Press, 1971.

Mosley, Leonard. *Power Play–Oil in the Middle East.* New York: Random House, 1973.

Penrose, Edith T. *The Large International Firm in Developing Countries: The International Petroleum Industry.* London: Allen and Unwin, 1968.

———. "The Changing Relations Between Companies and Governments," in *Economics of Energy: Readings on Environment, Resources and Markets.* Edited by Leslie Grayson. Princeton, N.J.: Darwin Press, 1975.

Philby, Harry. *Arabian Oil Ventures.* Washington: Middle East Institute, 1964.

Rand, Christopher T. *Making Democracy Safe for Oil: Oilmen and the Islamic East.* Boston: Little, Brown, 1975.

Rothberg, Burton G. *A Decision Theory Model of Eastern Hemisphere Oil Exploration.* Unpublished Doctoral Dissertation. Graduate School of Business Administration, Harvard University. Boston, 1975.

Sampson, Anthony. *The Seven Sisters.* New York: Viking Press, 1975.

Schatzl, Ludwig H. *Petroleum in Nigeria.* Oxford, England: Oxford University Press, 1969.

Shwadran, Benjamin. *The Middle East, Oil and the Great Powers,* (3rd. ed.). New York: J. Wiley and Sons, 1973.

Smith, David N. and Louis T. Wells. *Negotiating Third-World Mineral Agreements: Promises as Prologue.* Cambridge, Mass.: Ballinger, 1975.

Soldberg, Carl. *Oil Power.* New York: Mason/Charter, 1976.

Stobaugh, Robert B. "Evolution of Iranian Oil Policy, 1925-1975," in George Lenxzowski (ed.), *Iran Under the Pahlevis.* Stanford, California: Hoover Institution Press, 1978.

Stocking, George W. *The Oil Industry and the Competitive System: A Study in Waste.*

Westport, Conn.: Hyperion Press, 1976.

Sturgeon, James I. *Joint Ventures in the International Petroleum Industry.* Unpublished Doctoral Dissertation. Graduate College, University of Oklahoma, 1974.

Tanzer, Michael. *The Political Economy of International Oil and the Underdeveloped Countries.* Boston, Mass.: Beacon, 1969.

——. "The Energy Crisis: World Struggle for Power and Wealth," *The Monthly Review.* New York, 1974.

Teece, David T. "Vertical Integration and Vertical Divestiture in the U.S. Petroleum Industry." Stanford, California: Graduate School of Business Administration, Stanford University, 1976.

Tugendhat, Christopher and Adrian Hamilton. *Oil, The Biggest Business,* (rev. ed.). London: Eyre Methuen, 1975.

U.S. Congress, Senate. Special Committee Investigating Petroleum Resources. "The Independent Petroleum Company. Hearings," 79th Congress, 2nd. Session. Washington: U.S. Government Printing Office, 1946.

——, House. Select Committee on Small Business. "Effects of Foreign Oil Import on Independent Domestic Producers." 81st Congress, 1st Session. Washington: U.S. Government Printing Office, 1949-1950.

——, Joint Economic Committee. Subcommittee on Consumer Behavior. "The FEA and Competition in the Oil Industry: Hearings." (June 13, 1974), 93rd. Congress, 2nd. Session. Washington: U.S. Government Printing Office, 1974.

——, Senate, Committee on Finance. "Oil Company Profitability." (February 12, 1974). 93rd.Congress, 2nd. Session. Washington: U.S. Government Printing Office, 1974.

——, Senate, Committee on Finance. "1975 Profitability of Selected Major Oil Company Operations." (December 30, 1974). 94th Congress, 2nd. Session. Washington: U.S. Government Printing Office, 1976.

U.S. Congress, Committee on Ways and Means. "Background Readings on Energy Policy," 1975.

U.S. Office of International Energy Affairs. *The Relationship of Oil Companies and Foreign Governments.* Washington: Federal Energy Administration, 1975.

Vallenilla, Luis. *Oil: The Making of a New Economic Order: Venezuelan Oil and OPEC.* New York: McGraw-Hill, 1975.

Vernon, Raymond. "Foreign Enterprises and Developing Nations in the Raw Materials Industries." *The American Economic Review.* May, 1970.

——, et al. *The Oil Crisis.* New York: Norton, 1976.

——. *Sovereignty at Bay: The Multinational Spread of U.S. Enterprise.* New York: Basic Book, 1971.

——. *Storm Over the Multinationals: The Real Issues.* Cambridge, Mass.: Harvard University Press, 1977.

Wells, Louis T. "The Evolution of Concession Agreement." *Economic Development Report,* no. 117. Harvard Development Advisory Service. Cambridge, Mass., 1968.

Wilkins, Mira. "Multinational Oil Companies in South America in the 1920s." *Business History Review.* Autumn, 1974.

——. *The Maturing of Multinational Enterprise: Business Abroad from 1914 to 1970.* Cambridge, Mass.: Harvard Studies in Business History, Harvard University Press, 1974.

APPENDIX 2–DATA ABOUT THE INDEPENDENTS' INITIAL ENTRIES ABROAD AFTER WORLD WAR II

	Entry outside the U.S.				Entry outside North America			
	Year of Entry	Self-Sufficiency Ratio	Relative Size (%)[1]	Impts. Middle East/ U.S. Prod (%)	Year of Entry	Self-Sufficiency Ratio	Relative Size (%)	Impts. Middle East/ U.S. Prod (%)
Amerada	1948	P	2.1	1.2	1952	P	1.9	2.5
Ashland	1948	.22	1.9	1.2	1948	.22	1.9	1.2
Atlantic	1945	.53	13.9	.007	1945	.53	13.9	.007
Cities Service	1946	.46	21.9	.007	1946	.46	21.9	.007
Continental	1947	1.35	9.6	.026	1952	1.06	9.6	2.5
Getty	1948	P	.3	1.2	1949	P	.3	2.0
Marathon	1949	2.63	5.7	2.0	1952	2.46	5.4	2.5
Murphy	1951	P	na	1.7	1956	P	.3	4.0
Phillips	1944	1.04	12.6	0	1944	1.04	12.6	0
Pure Oil	1956	.44	6.7	4.0	1956	.44	6.7	4.0
Richfield	1946	.37	4.2	.007	1946	.37	4.2	.007
Signal	1948	P	.6	1.2	1948	P	.6	1.2
Sinclair	1945	.38	24.8	0	1945	.38	24.8	0
Skelly	1948	1.84	5.4	1.2	1958	1.31	3.3	5.1
St. Indiana	1948	.51	37.4	1.2	1948	.51	37.4	1.2
St. Ohio	1945	.28	7.6	0	1945	.28	7.6	0
Sun	1944	na	36.7	0	1956	.47	10.0	4.0
Sunray	1948	1.24	2.0	1.2	1957	.75	4.9	3.3
Superior	1947	P	1.5	.026	1947	P	1.5	.026
Tide Water	1949	.53	12.1	2.0	1956	.52	7.2	4.0
Union Oil	1947	.75	7.2	.026	1947	.75	7.2	.026

[1] Company's total income in % of Exxon's.
P = Producer only.
na = Not available.

SOURCE: Companies' *Annual Reports* for: year of entry, self-sufficiency ratio, size; U.S. Bureau of Mines' *Mineral Yearbooks* for: imports, U.S. production.

APPENDIX 3–VARIANTS TO THE BASE REGRESSION TEST

An alternative hypothesis was to assume that the independent oil companies reacted to world-wide imports rather than to the imports from the Middle East. When the latter started in 1946, the former were already high.

A second alternative hypothesis was to exclude Canada as a foreign country and therefore to consider only the entries outside North America. The rationale for that hypothesis is that many companies did not regard Canada in the late 1940s and the early 1950s as a foreign country.

A third alternative hypothesis was to suppose that the independents could have reacted in a given year to information (size and crude position, as well as measure of threat) dated from the previous year, (i.e., X1 t-1; X2 t-1; X3 t-1).

The eight combinations of the base-hypothesis are represented in a form of a decision-tree in Exhibit 1.

Exhibit 1
Regression Variants Pictured

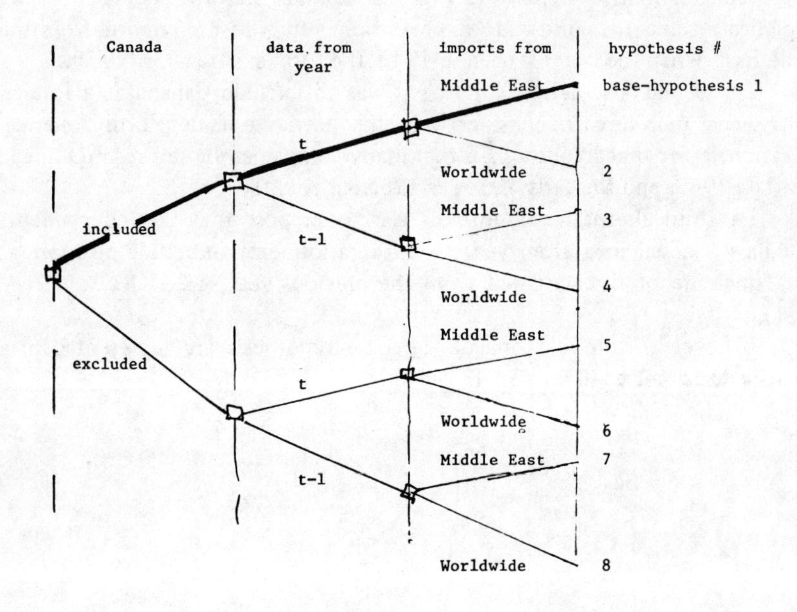

Exhibit 2
RESULTS OF EIGHT REGRESSION ANALYSES

Dependent variable = number of the year of entry abroad after 1945

	Regression # =	1	2	3	4	5	6	7	8
	Independent variables		Canada included				Canada excluded		
	Constant	1.33 (.99)*	-3.49 (.99)	1.90 (.99)	-5.95 (1.)	1.47 (.946)	-5.12 (1.)	1.39 (.914)	-4.14 (1.)
X_1	Self-sufficiency ratio (t)	.0013 (.751)	.0008 (.62)			-.0012 (.58)	.0055 (.99)		
	Self-sufficiency ratio (t-1)			.002 (.73)	.0034 (.99)			.009 (.91)	.002 (.72)
X_2	Relative size (t)	-.002 (.571)	-.004 (.59)			-.03 (.78)	-.003 (.60)		
	Relative size (t-1)			-.001 (.52)	.005 (.68)			-.003 (.53)	.005 (.592)
X_3	Imports from Middle East (t)	1.27 (1.)				2.48 (1.)			
	Imports from Middle East (t-1)			1.43 (.96)				2.45 (1.)	
	Worldwide imports (t)		.94 (1.)				1.21 (1.)		
	Worldwide imports (t-1)				1.56 (1.)				1.26 (1.)
	Number of Observations	11	13	11	11	12	15	12	13
	Correct signs?	yes	yes	yes	X_2 wrong	X_1 wrong	yes	yes	X_2 wrong
	Residual standard deviation	.308	.429	.750	.294	.923	.272	1.040	.577
	R^2	.918	.895	.513	.925	.958	.997	.947	.984
	Contribution to R^2, X_3	.684	.587	.297	.709	.744	.764	.735	.805
	X_2	.002	.000	.000	.000	.115	.116	.135	.132
	X_1	.232	.308	.216	.216	.100	.116	.078	.046
	Durbin-Watson test	1.77	3.19	1.38	1.56	1.86	2.63	1.64	2.06
	Serial correlation among independent variables? (level of confidence = 2.5%)	no	inconclusive	inconclusive	inconclusive	no	inconclusive	inconclusive	no

*The figures in parentheses give the probability for the sign to be correct.

APPENDIX 4
INDEPENDENTS' EXPERIENCES ABROAD BEFORE WORLD WAR II*

AMERADA was formed on June 4, 1919. By 1921, according to the Annual Report of this year it controlled over 8,000 acres of leasehold in Alberta (Canada), but no subsequent annual report will refer to it again. Indeed it was no accidental involvement abroad. Everette De Golyer was so bright a geologist that Sir Weetman (U.K.), for whom he worked in Mexican Eagle Co., chose him to organize a new oil company that would operate in the U.S. and perhaps in Canada. The company was named Amerada: a joining of America and Canada.[1] During 1926, it bought for cash a 31.7% interest in Esperenza Petroleum (a Delaware corporation organized in October 1926), which owned 335,000 acres in Venezuela, including 60,000 acres in the Lake Maracaibo district, and held purchase contracts covering about 370,000 acres.[2] However no exploration activity beyond surveying and geological work in 1934[3] was ever undertaken. In 1928 nevertheless Amerada increased its control over Esperenza up to 47.6% and until 1943 it spent between $50,000 and $85,000 each year just for maintaining the leaseholds. It appears that Amerada must have considerable confidence in its holdings in Venezuela for it has spent more than $1 million keeping Esperenza going.[4] Indeed Amerada's position was fairly summarized in Annual Report 1940: "No wells have been drilled, no drilling is contemplated at this time (although) a considerable portion of Esperenza's holding lies in the trend of development work now being undertaken by other companies."

ATLANTIC REFINING, was one product of the Standard Oil's dismemberment in 1911. Atlantic Lobos Oil Company was incorporated in September 1919 for the purpose of effecting a merger between the Port Lobos Petroleum Corp. and the Mexican properties of Atlantic's subsidiaries. By the end of 1925 Atlantic Lobos, affiliate at 50%, owned 100,000 acres of oil lands, two 20-mile long pipelines, tank storage for half-a-million barrels, a refinery with a capacity of 20,000 barrels a day. But the production was only of 1,160 barrels per day in 1925.

Between March 1922 and May 1925, Atlantic had set up five subsidiaries for marketing in Brazil, South Africa, Italy, Venezuela, and Colombia. All were supplied in petroleum products from the U.S. therefore providing "a large export business."[5] The Atlantic Union Oil Company, Ltd., was organized during the latter part of 1927 by Union Oil of California and Atlantic to market products in Australia and New Zealand. Marine terminals with a total storage capacity of 250,000 barrels were built at Sidney, Auckland, and Wellington at a cost of $4.1 million. 250 sales depots were set up; by the end of 1929 they numbered 800. In June 1933, however, Atlantic elected to sell its 50%

* The major part of the information reported in this Appendix was derived from the companies' *Annual Reports,* unless stated otherwise.

interest to Standard New Jersey at a profit of $1,319,600; Union did the same.

By 1930, Atlantic was "one of the largest manufacturers of lubricating oils in the world."[6] Marketing subsidiaries had been added in Germany, Portugal, Uruguay, and Middle East. By 1935 subsidiaries in Cuba, Belgium, Spain and North Africa were set up.

In 1936, because of the civil war, the line of communication with the Spanish subsidiary was completely interrupted and, therefore, a reserve of $247,000 established.[7] By 1938, for the first time, the foreign acreage hold was higher than the domestic acreage: 1,205,000 vs 1,107,000 acres. But while the foreign acreage portfolio was increased nearly eight times between 1937 and 1940 (from 927,000 to 7,246,000 acres located in Venezuela and Cuba), the foreign sales' share in total sales decreased from 29% in 1929 to 13% in 1939, and 8% in 1940. That means the centre of gravity of Atlantic's international strategy had shifted from the marketing-export end in the 1920s to the exploration phase by the end of the 1930s.

CITIES SERVICE, although an electric company, was engaged in oil business abroad as soon as 1918. In January a subsidiary was set up to operate leases in Mexico. By 1936, Cities Service accounted for almost 6% of the Mexican total production.[8] But all the properties were expropriated in March 1938. In June 1930 Cities Service and Standard New Jersey entered in joint venture for the development of over 400,000 acres in Venezuela. On the marketing side, Cities Service had been involved in export since 1922 with subsidiaries set up in Italy, Belgium and United Kingdom. In the early 1930s, Cities Service began to market petroleum products in Argentina under its brand name "Citex."

CONTINENTAL OIL, another former Jersey company, followed an international strategy far more limited in scope than Atlantic's. In 1920 it entered Mexico through two subsidiaries incorporated locally: Marland Oil Company of Mexico and Consolidated Oil Companies of Mexico. By 1929, it held approximately 25,000 acres of proven oil land in Mexico, which produced nearly 1,500 barrels per day. In 1926 it formed with Hudson's Bay Company[9] Hudson's Bay Marland Oil Co. (now Hudson's Bay Oil and Gas Co., Ltd.), which was given the privilege of exploring and developing company's oil rights in Canada. This joint venture, however, will stay inactive until after World War II. In 1940, less than 0.6% of sales of refined products were made abroad, but it was only export because the company had no foreign marketing facilities; and since expropriation of its Mexican properties in 1938 it had neither foreign production, nor foreign refining.

In April 1926 **MARATHON** (then Ohio Oil) purchased from Hoffer Oil Corp. 4 million acres in Mexico and incorporated Ohio-Mexico Oil Corp. under Mexican laws, retaining a 60% interest, which was increased up to 73% some years later. After having been unsuccessful for a long time, a gas reservoir was discovered in January 1931, the production of which was sold to Compania

Mexicana de Gas. Encouraged, the Ohio-Mexico acquired 29,000 additional acres where it drilled three new gas wells that same year.

From the Mexican adventure Marathon learned a much-needed lesson: foreign operations cannot be easily successful. Somebody must be at the scene to meet the myriad problems of operating in a foreign language and in a different culture. Indeed the Mexican subsidiary had, for example, only seven employees in 1932. In addition the first Mexican properties were expensive to acquire and dear to own. At the beginning they yielded only dry holes. Even when, by 1937, about 20 million cubic feet daily of gas were sold, the revenues were insufficient to justify further development.[10] It was not saved, however, by the nationalization of 1938. After five more years of financial troubles, Marathon sold the properties and retired from Mexico.

PURE OIL followed an international strategy similar to Amerada's. By 1926, it owned 150,000 acres in Lake Maracaibo through a 75% interest in Orinoco Oil; 25% being owned by Venezuelan interests. By 1930 the acreage amounted to more than 300,000 acres "on which the exploration work had been completed."[11] The subsidiary was mentioned for the first time in Pure's Annual Report only in 1934, and yet only in footnote. From then on it was referred to as "mainly undeveloped concessions." Nevertheless by the end of 1944 the investment in that Venezuelan subsidiary amounted to $5,225,209 representing 75% of outstanding capital stock and annual advances.[12]

RICHFIELD'S experience abroad was mainly on the marketing end, but very limited. In 1926 Rio Grande–which became part of the heritage of Richfield–launched its first and only venture into Mexico. That year the company completed a small topping plant below the border in Nogales, mainly to produce kerosene for the lamps and stoves of Mexico. From the beginning the business prospered. Within six months almost all the plant's capacity was being marketed. Rio Grande began thinking of expanding and asked a retired army general and former president of Mexico to become general agent for the west coast. The negotiations went well until Rio Grande was outbid by Standard Oil of California. Shortly afterwards import duty on kerosene was rescinded, allowing Socal to export from the U.S. instead of refining in Mexico and leaving Rio Grande without tariff protection. At that moment there was nothing left for Rio Grande to do but salvage what it could out of its Mexican investments. That was accomplished by selling the plant and the plant site in Nogales to Socal for construction of its bulk plant. "Rio Grande had learned something about investing where political protection may be withdrawn without notice."[13]

Richfield's foreign strategy as Richfield Company never went beyond export. The company had a contract to supply refined products–mostly gasoline–to Australian Motorists Petrol. Ltd. That firm had been awarded a 10% share of the total automotive-fuel market of the Australian Gasoline cartel. It was a sizable market which Richfield had been supplying for years before the

U.S. entered the war.[14]

SINCLAIR, before World War II, had one of the most consistent international strategies among the independents. The objectives were deliberately set right at the beginning. Harry F. Sinclair envisioned in his first proposal back in 1916 an organization "engaged in all branches of the petroleum industry and *international in its scope of operation.*"[15] Indeed, by 1925, the company had large concessions in Costa Rica, Panama, Mexico, and Angola.[16] This last country was Sinclair's first entry outside the Western Hemisphere, and it occurred in 1916. The Compania de Petroleo de Angola had obtained an oil concession on forty million acres of land in Angola. Sinclair had a half stock interest in this company along with Belgian and Portuguese groups and was operator. By 1929 twenty wells had been drilled, but with no oil in commerical quantities. Some years later the concession was relinquished.[17]

In 1921, Sinclair began to seek a concession in Persia. According to one of its executives, Sinclair started to negotiate with the Iranian authorities "following up a suggestion of the U.S. Secretary of Commerce," who was then Herbert H. Hoover.[18] Sinclair happened to be directly competing with Standard Oil of New Jersey for the same concession. In fact, at that time, that is seven years before the Red Line Agreement, which would close the doors of the Middle East behind the 'majors,' Standard New Jersey was struggling itself with the British companies for the control of a piece of Middle East oil.

Although Sinclair eventually got the concession in December 1923, it withdrew quickly. The reason is that Sinclair hoped to use the concession in north Persia in connection with its business in the Soviet Union.

In 1923, Sinclair tried to do business with the Soviet Union. The Russian government had confiscated all oil properties owned by foreign companies, but Barnsdall International Oil Company managed to obtain a number of concessions in the Baku area under contract. Sinclair was to take an interest with Barnsdall. Two wells were drilled, but they were then taken over by the Russians. In an executive's opinion: "We had simply made the mistake of believing that an engagement by the Russians to permit us to produce at Baku would be honored."[19] Accordingly, Sinclair notified the Persian government early in 1925 that the attitude of Soviet Union prevented it from pursuing further its concession in northern Persia.

More serious was the venture into Venezuela. In December 1928 Sinclair acquired a majority interest in the Venezuelan Petroleum Company which had concessions in Venezuela, Colombia, Panama. In Venezuela, concessions in Lake Maracaibo were developed by Gulf Oil on a royalty basis and by Standard Oil of California on an interest basis. Sinclair had been in Venezuela as early as 1922, however, but only as a marketer. Other marketing subsidiaries were set up in 1919 in Mexico and Cuba, in 1920 in Belgium, in 1923 in United Kingdom. By 1935 the geographical break-up of consolidated gross operating income was 85% in the U.S., 5% in export, 4% in Europe, 3% in Cuba and 3% in Mexico. From 1937, Sinclair stepped up exploration in Venezuela. It was good timing for in 1938 the Mexican properties were nationalized. But it was only in 1941 that Sinclair brought in its first successful well in Venezuela. By then, with Hitler's rise to power, it had sold its European marketing subsidiaries for cash and barter goods,[20] and terminals, bulk plants and other distribution facilities were mentioned in Cuba only.[21]

STANDARD OF INDIANA. In 1925 Colonel Stewart, Chairman of Standard Indiana, accomplished in large measure his ambition to make his company a major crude oil producer through the acquisition of certain of Edward L. Doheny's interests in the Pan American Petroleum and Transport Company, then "one of the largest crude oil producers in the world."[22] The mechanism by which the purchase was executed is somewhat complicated and needs not to be described in detail. Suffice it to say that Standard acquired control with 50% of the voting stock for $37,575,000 in what was called "one of the greatest and most sensational deals of oil history."[23] Indeed the deal, which was equivalent to a merger of Pan American (excluding its California properties) with Standard of Indiana, could be qualified as the largest oil consolidation in the history of the industry with combined assets valued at nearly $584 million.

It has been suggested that Colonel Stewart was "among those who believed that the U.S. would soon have to turn to foreign sources for a large share of its crude oil supplies."[24] Pan American's control fulfilled also this objective. The deal included about 1.4 million acres of oil land in Mexico with a current production of about 150,000 barrels a day equally divided between heavy and light crude. The Mexican field was considered at the time one of the most important and valuable in the world.[25] Pan American interest had the largest refinery in Mexico at Tampico with a capacity of 130,000 barrels a day, 640 miles of pipeline, landing terminals, and a railroad.

An extremely important feature of the Pan American acquisition was that it gave Standard Indiana a 50% interest in the British Equatorial Oil Company. The concessions covered 3 million acres around and under the shallow waters of Lake Maracaibo, with current production of about 45,000 barrels per day. A refinery at La Sabina in Venezuela manufactured products for local consumption. The British Mexican Petroleum Company was also one of the leading distributors of fuel oil in the world. It had marketing outlets throughout the United Kingdom and operated a fleet of tankers for making shipments to many foreign countries. In addition British Mexican was under contract to purchase several million barrels of crude oil and gasoline a year from the Mexican Petroleum Company, Pan American's principal subsidiary.

To transport these petroleum liquids, Pan American had one of the largest tanker fleets in the world. It was valued at $10 million and had a carrying capacity of approximately 3 million barrels. In addition to marketing facilities in the U.S., Pan American's subsidiaries had established a number of bulk stations for marketing fuel oil and a few service stations in South America at Cristobal (Panama Canal Zone), Buenos Aires (Argentina), Montevideo (Uruguay), and in several cities in Brazil.

Acquisition of all these foreign properties of Pan American placed "under control of Colonel Stewart sufficient properties to give him the position of almost unequaled influence in the oil world."[26] The Standard Indiana, according to MacLean and Haig,[27] was also placed in a position to consolidate its marketing outlet to foreign countries in competition with Standard New Jersey and other interests if it so desired. But more important it gave Standard Indiana the crude oil reserves necessary to support its refining and marketing operations. In 1925 Standard Indiana had a refinery capacity of 160,000 barrels per day whereas the producing properties acquired from Pan American had a crude oil output of 140,000 barrels per day.

After the acquisition Colonel Stewart expressed the opinion that Standard Indiana had solved its crude oil problems for a long time to come.[28] The prediction would hold for just seven years.

The Decision To Sell All Foreign Properties

During the following years, Standard Indiana invested management efforts and money in consolidating and expanding its foreign assets. Where others had failed before, in 1927 Pan American was one of a group of five U.S. oil companies which gained a quarter interest in the Iraq Petroleum Company. Late in 1927, the decision was made to build a refinery at Aruba (Dutch West Indies) with a capacity of 115,000 barrels per day in order to process Venezuelan crude oil. In 1929 another refinery was under construction at Hamburg (Germany) for the production of 5,000 barrels a day of asphalt.[29] Supplies were to come from Mexico; a German subsidiary was organized to market the asphalt in Central Europe, another one in U.K. By the end of 1930, marketing organizations abroad owned or controlled 14 ocean terminals, 59 interior bulk plants and 70 service stations. During the same year, statistics of manufacturing included for the first time news of refinery at Aruba and a full year's operation of the Asphalt plant at Hamburg.[30] Last, in 1931 an increase in marketing facilities in foreign countries was gained through the purchase of stock interest in the Petroleum Storage & Finance Corporation, Ltd. of Manchester (U.K.)[31]

Then, quite unexpectedly, in a complete reversal of strategy, in May 1932 Standard Indiana sold all its foreign properties to Standard (New Jersey). Several rationales have been offered for this "outstanding event in the history of Standard Indiana."[32] According to the company's management the decision was prompted by an impending import tax on crude oil and refined products. Indeed the Pan American crude oil and refined products had so far been used primarily to supply the Standard and Pan American refining and marketing facilities in the U.S. As the Standard's directors saw it they had to choose between two alternatives: 1) to spend millions of dollars to develop a foreign distribution system in order to provide an outlet for the Venezuelan and Mexican crude and the refined products from Aruba; or 2) to sell the foreign properties.

As a matter of fact Standard's management had practically no choice. Since 1929 Pan American had been spending millions to build up its foreign marketing organization. To firmly establish a place for itself in overseas markets would require an additional outlay of money. "It could be done, but it would not be easy."[33] In fact it would have been very difficult to spend this money in the midst of the Depression when oil prices were low and company's sales were at their lowest in many years. With a tariff, these expenditures would be even greater. In addition, Standard Indiana would be faced with the task of developing new sources of supply for its domestic operations. It appeared that the simultaneous undertaking of these two projects would pose insurmountable financial problems, and the sale to the Jersey Company was therefore adopted as a practical alternative solution.[34]

The irony is that Standard Indiana participated in the building up of the situation which led it to disinvest from abroad. Pan American, Shell, Gulf, Tidewater, and Standard of New Jersey were importing crude oil and gasoline in large quantities from Venezuela. With more crude than the market could possibly absorb, some producers in the Mid-Continent and Gulf Coast areas particularly began agitating in 1929 for the imposition of a tariff duty. It was unsuccessful. But as the depression deepened, domestic producers increased their pressure for a tariff. Pan American was the principal target. In 1931, Pan American set up an office in Washington to lobby against the passage of any tariff; it was successful. To prevent Congress from taking action, Standard New Jersey and Gulf proposed to Standard Indiana a voluntary reduction of 50% in Venezuelan imports, but the latter was against it. When Indiana decided to follow the others' move to reduce imports, it was almost too late.

This indecision is quite significant. According to McLean and Haig: "It seems highly probable that the decision to make the sale also reflected the inherent conservation of the Seubert administration as contrasted with the Stewart administration."[35] Giddens, the company's historian, shares the same view: "Standard's directors, lacking the fighting spirit of Colonel Stewart and not being experienced foreign oil operators, decided to sell Pan American's foreign properties."[36] Had they had a consistent international strategy, they would probably have not sold all the foreign assets; some being well integrated abroad like the asphalt business in Europe, some worth more than the book value they were sold at like the estimated Venezuelan reserves of 550 million barrels.

The management said "the deal materially improved Standard Indiana's position in relation to the world trade in oil"[37] because, in addition to $50 million in cash, it became Jersey's largest stockholder with 7% of the outstanding stock. It remains that Standard New Jersey had acquired control over the most valuable crude oil properties in the Western Hemisphere.[38] Standard Indiana had lost its place among the majors: "This transaction eliminated a potentially strong competitor in foreign markets."[39] Had it stuck to Colonel Stewart's international strategy, Standard Indiana would be now "the second U.S. major."[40]

SUN OIL went abroad both to explore and to market, but mainly to do the latter. By 1925 its products (mainly lubricating oils) were distributed in France, Italy, Belgium, United Kingdom (since 1909), Germany and Holland. Five years later it had sales offices and distributing branches in Canada, Mexico, eight countries of South America, the West Indies, Cuba, eight countries of Europe, Australia and India. It had even an export business towards China. In March 1930 Sun planned a chain of service stations in Canada. By the end of 1940 out of $3,149,623 invested abroad (2.2% of total assets), 87.4% were in Canada. Sun's experience in foreign exploration was limited to Venezuela where it shared since 1930 with Standard New Jersey 1,750,000 acres in the Lake Maracaibo.

By virtue of its extraordinarily favorable position on New York harbor **TIDE WATER** had since the late 1880s sold a very large percentage of its refinery's output to the export trade. In 1919 an Export Department was created with very definite plans for marketing in foreign lands.[41] By 1930 a subsidiary, Associated Oil Company, had service stations in the Philippines, sold refined products to Australia, New Zealand, Indo-China, Japan, and fuel oil on the west coast of South America. Another subsidiary, Tide Water Oil Company, was engaged in marketing of refined products in Argentina, Uruguay, Paraguay, Brazil, and controlled subsidiaries in Italy, Germany, France and U.K. At this time the only interest in exploration-production abroad was in Mexico.[42] In 1934 the Annual Report mentioned for the first time a 50% interest in the Mitsubishi Oil Company, Ltd., which operated a refinery near Tokyo. By the end of 1939 the book value of foreign investments amounted to $2,036,312, of which 73% were located in Japan.[43]

UNION OIL OF CALIFORNIA, as Tide Water, was abroad through export trade even before World War I. *Annual Report 1907,* for example, mentions marketing in Chile, Panama and Guatemala. In 1918 the Union Oil Company of Mexico was formed to develop a 16,000 acre lease granted by the Mexican government. In 1920 two discovery wells in Mexico produced 4.6 million barrels of oil in six months. From then and until the Depression, Mexico was to supply the South American marketing organization. Late in 1921 a Canadian subsidiary was formed to produce refined oil at a refinery in Vancouver and operate service stations in Western Canada. In 1923 the portfolio of foreign concessions included Mexico and Colombia. But late in 1926 Union took a 50% interest at a cost of $3.5 million in the Pantepec Oil Company of Venezuela. As mentioned earlier Union pulled out of Pantepec in 1931, but kept in Venezuela 260,000 acres on its own and 40,000 additional acres developed under royalty interest by Standard New Jersey. By the end of the 1930s Union had direct business abroad only in Canada, Panama, and Chile. But it had a large export business towards Europe, Orient and South America. Excluding Canada, exports represented 13% of dollar sales in 1938.

NOTES

[1] *Fortune,*"Amerada plays them close to the chest." January 1946.

[2] Amerada's *Annual Report 1926.*

[3] Ibid., *Annual Report 1934.*

[4] *Fortune,* op. cit., January 1946.

[5] *Moody's Industrial 1926.*

[6] *Moody's Industrial 1930.*

[7] Atlantic Refining's *Annual Report 1936.*

[8] Wilkins, M., op. cit., p. 226.

[9] Incorporated in England in May 1670.

[10] Hartzell, Spence, *Portrait in Oil–How the Ohio Oil Company Grew to Become Marathon* (New York: Mc-Graw-Hill, 1962), p. 317.

[11] *Moody's Industrial 1930.*

[12] Pure Oil's *Annual Report 1944.*

[13] Charles S. Jones, *From the Rio Grande to the Arctic–The Story of the Richfield Oil Corporation* (Norman, Oklahoma: University of Oklahoma Press, 1972), p. 36.

[14] Ibid., p. 219.

[15] Quoted in Sinclair's *Annual Report 1922;* italics added.

[16] *Moody's Industrial 1936.*

[17] Connelly, W., *The Oil Business As I Saw It–Half A Century With Sinclair* (Norman, Oklahoma: University of Oklahoma Press, 1954), pp. 110, 120.

[18] Cited by B. Shwadran, *The Middle East, Oil and the Great Powers,* p. 86, n. 23.

[19] Connelly, *The Oil Business,* p. 100.

[20] At a profit of about $1 million.

[21] *Moody's Industrial 1942.*

[22] Paul H. Giddens, *Standard Oil Company (Indiana)* (New York: Appleton-Century-Crofts, 1935), p. 240.

[23] Giddens, *Standard,* p. 240.

[24] McLean, J. and R. Haig, *The Growth of Integrated Oil Companies,* p. 259.

[25] Giddens, *Standard,* p. 242.

[26] *National Petroleum News,* April 1, 1925, p. 33; quoted in McLean and Haig, *Growth of Integrated Oil Companies.*

[27] McLean and Haig, op. cit., p. 260.

[28] Quoted in McLean and Haig, op. cit.

[29] *Moody's Industrial 1930.*

[30] Standard Indiana's *Annual Report 1930.*

[31] Ibid., *Annual Report 1931.*

[32] Giddens, *Standard,* p. 489.

[33] Ibid., p. 489.

[34] McLean and Haig, *Growth of Integrated Oil Companies,* p. 262.

[35] Ibid., p. 262.

[36] Giddens, *Standard,* p. 489.

[37] *Oil and Gas Journal,* May 12, 1932, p. 22.

[38] Giddens, *Standard,* p. 293.

[39] Larson, H., E. Knowlton and C. Popple. *History of Standard Oil (New Jersey)–New Horizons, 1927-1950* (New York: Harper & Row, 1971), p. 311.

[40] Interview by author in Standard of Indiana.

[41] Buente, R., "Autobiography of an Oil Company," *Tide Water Topics,* Nov.-Dec. 1923, p. 60.

[42] *Moody's Industrial 1930.*

[43] *Moody's Industrial 1941.*

APPENDIX 5
INTERACTION BETWEEN INDEPENDENTS IN FOREIGN EXPLORATION

Operating in an oligopolistic industry, one would expect the independents to behave accordingly, that is, to react defensively to any competitor's move that could threaten its own position. I hypothesized that the independents' entries into foreign exploration were bunched in time and toward the same countries. The test developed by Knickerbocker–called entry concentration index (ECI)–was used.[1] Knickerbocker defined the concept as: "a quantitative measure of the extent of oligopolistic reaction within a given industry."[2] A simplified example of the construction of an ECI is presented in Exhibit 1. Defined as such, the ECI gives the maximum propensity among the independents to initiate exploration within a three-year time span in the same country or set of countries.

Across industries, Knickerbocker found evidence that entries by U.S. firms into foreign production were bunched in time, more so than could be expected by chance. However, as far as this research is concerned, his test dealt with only one phase of the petroleum industry: refining. In addition, the data base was the one gathered by the Harvard Multinational Project where the sample of U.S. oil companies was a mix of all the majors and of the four largest independents (Cities Service, Continental, Phillips, Standard Oil of Indiana). There was, therefore, a need to replicate Knickerbocker's test.

All entries into exploration abroad by the twenty-two independents between 1945 and 1976 were recorded. But for the computation of the global ECI only the countries with three or more entries were taken into account on the argument that an ECI of 1 based on one or two entries is meaningless. A distinction was then made between first entry by a given independent in a new country and its successive reentries in the same country. Under these conventions, 456 first entries and 227 re-entries distributed among 55 countries were recorded.

For the purpose of this study, only three-year clusters were taken into consideration while Knickerbocker dealt also with 5-year and 7-year clusters. The rationale for the choice of a tight test of "follow-the-leader" behavior is that it is not necessary to go beyond a geological-political risk analysis for a decision to go exploring abroad while an industrial investment requires financial and marketing feasibility study. In other words, one would expect a quicker reaction from oil companies to a geological opportunity or to a threat by a rival.

The test's results are summarized in Exhibits 2 and 3.

By comparison with Knickerbocker's test results, which gave an ECI of .46 for 3-year clusters in the case of 187 U.S. multinationals, I chose an ECI of .5 as lower bound for evidence of follow-the-leader patterns.

Exhibit I
SIMPLIFIED EXAMPLE OF THE CONSTRUCTION OF AN ECI

	Year										Maximum # of Independents Going to Explore Abroad Any 3 Consecutive Years	Total # of Independents Going to Explore Abroad
	1	2	3	4	5	6	7	8	9	10		
Entry into Country A												
# of Independents Buying Exploration Concessions in Year Indicated	3	5	5	2	1	2	2	1	1	1	13	23
Entry into Country B												
# of Independents Buying Exploration Concessions in Year Indicated	3	1	1	3	4	2	1	1	1	1	9	18
Total: Countries A & B											22	41

$$\text{Three-year Entry Concentration Index} = \frac{\text{Maximum 3 Consecutive Years}}{\text{Total of Independents Going To Explore Abroad}}$$

$$= \frac{22}{41} = .538$$

(Adapted from Knickerbocker, thesis, op. cit., p. 46.)

Exhibit 2 – ENTRY CONCENTRATION INDEXES IN FOREIGN EXPLORATION

	ECI
First entry worldwide	.570
First entry and re-entries worldwide	.482
First entry into:	
Canada	.500
Rest of Western Hemisphere	.429
Total Western Hemisphere	.329
Eastern Hemisphere	.286
Europe	.386

From Exhibit 2, it appears that, on a worldwide basis since World War II, the independents tended to match the rivals' moves in the search for new oil reserves: in average four out of seven new entries (57%) occurred within clusters of three years. However, it is also obvious that it was not made on a hemisphere basis although several independents had their exploration department divided between Western and Eastern Hemispheres for a long time.

A substantial majority of forty-one foreign countries (75%) triggered significant oligopolistic reactions (ECI above 50%). This result supports the hypothesis that the independents bunched their entries on a country-by-country basis.

Over the period studied, these bunchings in 3-year clusters were distributed in the following way: in the 1940s, 5%; in the 1950s, 22%; in the 1960s, 39%; in the 1970s, 34%. It seems, therefore, that it is the second half of the period, when the independents were no more exploring abroad in response to threat by imports, but in order to optimize their geographical portfolio of oil reserves, that they display a higher propensity to watch and match the competitors.

Among the few countries which revealed no follow-the-leader pattern, the most unexpected ones–because they were the center of intense petroleum activity–are Argentina, Libya and Nigeria. For what I was investigating is the pattern of follow-the-leader into producing fields, a more appropriate way of defining entry would be to consider parts of countries or oil reserves instead of the entire countries. If the analysis could be made on this basis, the lower ECIs associated with Canada, UK, North Sea, Iran, and Indonesia and the three countries mentioned above might rise. Gathering data on a producing field basis, however, would require some kind of geological criterion that could be added in further research developments.

The fact that the follow-the-leader hypothesis is eventually accepted is partly due to the oil companies' propensity to go exploring through joint-ventures. According to the companies it is to share the financial risk. But it is also a form of defensive action in order to hedge the risk of not being present where a new big field might be discovered. However, not all the independents

Exhibit 3
ECI BY COUNTRY

ECI in % / # of entries between:	3-6	7-10	11-14	15-19	20-23
80+	Germany onshore German North Sea UK onshore Greenland Portugal Kenya Ghana Bangladesh Viet-Nam	Angola	Netherlands North Sea		
66-79	Mexico Ecuador Somali Thailand Brunei	Philippines	Guatemala Spanish Sahara	Peru	
50-65	Honduras Nicaragua Paraguay Mozambique Saudi Arabia Abu Dhabi	Ireland Italy Spain Norway North Sea Algeria Pakistan Papua-New Guinea	Egypt Australia	Bolivia Iran Indonesia	Canada Venezuela UK North Sea

went into a country through the same joint-venture, nor in the same year. In other words, whatever the tactic they chose, the fact remains that more than one independent happens to converge on the same geographical area in a relatively short period of time.

NOTES

[1] Frederick T. Knickerbocker, "Oligopolistic Reaction and Multinational Enterprise," op. cit.

[2] Ibid., p. 44.

INDEX

Aharoni, Yair, 51
Amerada, 20, 44, 75, 79, 82-88, 133
Amerada-Hess, 88-90
Andrews, Kenneth, 51
Aramco, 18, 22
Ashland, 78, 82
Atlantic, 24, 67, 100, 104, 133
Atlantic-Richfield, 102

Barriers to entry, 19, 53
Bower, Joseph, 51
Brooke and Remmers, 52
Burmah Oil, 111

Chandler, Alfred D., 51, 52, 121
Chenery, Hollis B, 20, 22. *See* Mikesell
Church Committee, 19
Cities Service, 24, 67, 96, 134
Compagnie Française des Pétroles, 111
Company's top executive, 19
Conorada, 84f
Continental Oil, 67, 84, 86, 88, 94, 97, 98, 99, 100, 102, 104, 134
Crude long company, 27, 40, 70, 117, 119
Crude short company, 27, 40, 70, 71, 80, 94, 117

DeChazeau and Kahn, 55
DeGolyer, E., 20, 133
Demand for oil, 20

Experience abroad, 28, 33, 41, 45, 63, 71-72
Exploration: cost of, 22, 27, 45; effort over time, 79-80, 94-97; geographical pattern, 80-81, 93-94, 111, 117
Explorer-producers, 67-68, 75

Fohs, Julius, 21
Follow-the-leader, 13, 15, 45, 120
Foreign tax credit, 18-19
Frankel, Paul, 19

Gabriel, Georg, 20
Getty Oil, 18, 44, 67, 96-97
Getty, Paul, 19
Gulf Oil, 23-24, 108

Hammer, Armand, 19
Hunt, Michael, 54
Hunt Oil, 4
Hymer, Stephen, 14, 53

Imports: from the Middle East, 23-24, 27, 33-36, 45, 108, 118; controls or quota, 75-79, 81, 93, 97-98, 110, 118-119
Independent: definition, 1-5
Initial decision to go abroad, 1, 14
Iranian Consortium, 33, 67, 109, 111

Jacobson, A., 83, 87
Jacoby, Neil, 4, 15, 21-22
Joint-venture, 20, 81, 84

Kerr-McGee, 4
Knickerbocker, Frederick T., 14, 145

Literature on: foreign direct investment, 13-15; industrial organization, 53-54; international business strategy, 52-53; petroleum, 15; strategy, 51-52
Lumpiness of foreign investment, 20, 22, 45, 70, 107, 117

Marathon Oil, 67, 84, 88, 99, 134
Marketing abroad, 99-102, 108
McCullough, E.H., 87
McLean and Haig, 120
Mikesell, Raymond, 20, 22
Mobil, 87, 110
Mosley, Leonard, 19
Multinationalization through verticalization, 55, 99
Murphy Oil, 44

Occidental, 44, 67, 97, 99, 102
Ohio Oil. *See* Marathon
Oil Crisis, 82, 90, 102, 118-119
Oil policy of host government, 19
Oil reserves, company's, 22
Oligopolistic reaction, 14, 22, 121

Philips Petroleum, 44, 75, 96-97, 102
Pogue, Joseph, 20
Porter, Michael, 54
Portfolio theory, 14
Product life cycle model, 14
Pure Oil, 44, 71, 78-79, 81, 135

Refining abroad, 98-99, 110
Research: methodology, 6, 8, 54-59; objectives, 1; outline, 8-9
Richfield, 66-67, 78, 135
Royal Dutch-Shell, 86
Rumelt, Richard, 52

Sample, 1-6
Sampson, Anthony, 19
Scott, Bruce, 51-52, 54
Self-sufficiency, 33, 40-41, 75, 80, 94, 103-104, 117
Signal, 4, 44, 67, 96, 98-99, 102
Sinclair, 8, 24, 44, 67, 97, 102, 111, 136
Size, company's, 20, 28, 33, 36-40, 45, 70, 97, 117, 119
Skelly Oil, 75, 81
Socony-Vacuum, 23, 24
Soldberg, Carl, 18
Stages of development, 54
Standard (New Jersey), 23, 24, 67, 134, 140
Standard Indiana, 24, 67, 96, 102, 137-140
Standard Oil of California, 23, 135
Standard Oil of Ohio (Sohio), 4, 5, 44, 67, 75, 78, 79, 80, 82
Stobaugh, Robert B., 14
Stoford and Wells, 53-54
Strategic group, 53, 66, 67, 120
Strategy: definition, 51; international, 52-54; over time, 1, 52, 79, 104, 118, 120
Sun Oil, 44, 66, 67, 96, 98, 104, 140
Sunray, 8, 78
Superior Oil, 44, 75, 81

Tax incentives, 18-19
Tenneco, 4
Texaco, 75
Texas Gulf, 4
Threat by the majors, 23, 27, 33-36, 45, 107
Tide Water, 8, 24, 67, 140
Totally vertically integrated companies, 67-68, 93
Transportation cost, 24

Union Oil, 67, 96, 102, 133, 134, 140

Vallenilla, Luis, 20
Vernon, Raymond, 4, 13, 14, 20, 23, 55
Vertical integration, 27, 54-55, 66, 71, 87, 97f, 103-104, 111, 117

Wells, Louis T., 4. *See* Stopford and Wells
Wilkins, Mira, 4, 54

DATE